Nonlinear Viscoelastic Solids

Nonlinear Viscoelastic Solids

F. J. LOCKETT

Division of Materials
Applications,
National Physical
Laboratory,
Teddington, Middlesex,
England

1972

Academic Press London · New York

ACADEMIC PRESS INC. (LONDON) LTD.
24/28 Oval Road,
London NW1

United States Edition published by
ACADEMIC PRESS INC.
111 Fifth Avenue
New York, New York 10003

Library of Congress Catalog Card Number: 72–84349
ISBN: 0-12.454350.2

PRINTED IN GREAT BRITAIN BY

William Clowes & Sons Ltd., London, Colchester and Beccles

Preface

The mechanical behaviour of materials can be classified under a number of headings which depend on the type of material considered and on the applications for which the material is intended. When a solid is used in a situation which may involve significant loading and deformation, it is essential that adequate design should be included in the production of the article or component. This may be necessary to ensure efficient material utilization, to prevent unacceptable deformation during service, or to avoid failure of the component. In part these design requirements can be met by providing methods for stress and deformation analysis for the relevant material behaviour. Essentially, this requires some form of theory to model the behaviour, experimental techniques for measurement of material parameters, and methods for performing the calculations relevant to a particular application. Obviously the degree of sophistication employed will be governed by the particular application, and the methods of calculation may involve detailed mathematical or numerical analysis, use of simplified engineering theories, or the use of data sheets. Whatever methods are employed, the theoretical basis and experimental data should have a corresponding degree of sophistication. This book is concerned mainly with a class of situations in which a relatively high degree of accuracy is required.

When the relevant material behaviour can be adequately modelled by either the linear theory of elasticity or one of the theories of plasticity, appropriate theories are available and are well understood, and experimental methods for determining material parameters are well established. In these circumstances the solution of a design problem reduces to a calculation relevant to the given situation using an established theory and known data. When the behaviour is linear viscoelastic, mathematical theory and experimental methods are available, but the solution of individual problems becomes more difficult. In addition, the development and documentation of engineering theories have received less attention than the corresponding elasticity or plasticity theories, and are generally less well understood by the design engineer. Similar remarks apply to the development and use of design methods for nonlinear elastic behaviour.

Design procedures for nonlinear viscoelastic behaviour are at an earlier stage of development than those mentioned above. In this case there are no established theories which are acceptable from the point of view of practical utilization and, for this reason, little progress has been made in the collection of design data and development of practical design procedures. This is not to say that there does not exist a highly developed mathematical theory of nonlinear viscoelasticity. However, this has been developed as an exercise in theoretical mechanics and has had little practical impact. In general the complexity of the theories is such that the solution of stress-analysis problems would require extensive numerical computation, and experimental methods for providing the relevant material data are generally not understood even in principle. In part this situation is due to the fact that theoreticians and experimentalists have often worked in isolation, whereas an interdisciplinary approach is essential to success in this complex subject. This book is intended to provide a stimulus for interdisciplinary work by providing a common ground on which theoreticians and experimentalists can appreciate each others problems and assess the practical benefits to be gained through collaboration.

This book is devoted entirely to the associated problems of developing adequate theories for describing viscoelastic behaviour, and designing experimental tests for measuring material data. Solution of these problems is an essential prerequisite to the development of design procedures and the solution of individual stress-analysis problems. Emphasis is placed on general principles rather than on rigorous mathematical analysis or details of experimental techniques. In particular, the theories which are discussed are confined to those which have been used to model experimental data, or appear to be suitable for further consideration. The subject of the book is such that it has not been possible to write a definitive text which answers general questions, or even places proposed methods in an order of merit. Instead it provides a detailed exposition of the present state of the art which can be used as a starting point for further interdisciplinary research.

The restrictions which are placed on the material behaviour for the purposes of this book are discussed in Chapter 1. In general, the discussion is confined to the macroscopic behaviour of isotropic solids under isothermal conditions, although some discussion of non-isothermal behaviour is included. Plasticity associated with a yield stress is not considered, but the types of behaviour which are considered are appropriate to many polymers, such as plastics and rubbers, in in-service conditions. A general introduction to the subject is given in Chapter 1, and the theory of linear viscoelasticity and its associated experimental methods are treated in Chapters 2 and 3. These three chapters provide an adequate basis for understanding the discussion of nonlinear behaviour contained in the remainder of the book. Chapter 4 gives an outline of relevant contributions to the nonlinear theory of viscoelasticity, and Chapter 5 deals with the determination of material properties for one-dimensional behaviour and with the analysis of available experimental data. Three-dimensional behaviour is considered in Chapter 6 and it is shown that present theories, which claim to model relatively general behaviour, are too complex to be used in practice. The last three chapters deal with various types of simplified theory which have been motivated either by mathematical, physical or empirical assumptions. Much of this work is supported by comparison with experimental data, and in many cases data have been modelled adequately by relatively simple theories in restricted circumstances. The results described undoubtedly contain the basis for further advances in the subject, but present evidence is inconclusive for identifying the most promising areas with any degree of certainty.

The subject of the book, and the level of knowledge required of the reader, are such that it can be used by research workers and as a graduate-level text. Since it deals with the interface between the theoretical and experimental approaches, the book will be of use to research workers in both disciplines. It will help the theoretician to view his work in a practical perspective and to identify the objectives to which further work should be aimed, and the experimental worker will be provided with some insight into the achievements, limitations and possibilities of rational mechanics. Although the presentation relies heavily on mathematical notation, this has been kept as simple as possible and a high level of mathematical ability is not required. For this reason the book can be used as a supplementary text for courses in theoretical or engineering mechanics which deal with viscoelastic behaviour. In general the book is pragmatic rather than scholarly, and the specialist, be he theoretician or experimentalist, may feel that in certain respects it lacks rigour or complete detail. This is deliberate, and is a necessary consequence of writing a book for an interdisciplinary audience.

It is a pleasure to acknowledge the generous assistance given by a number of friends and colleagues. In particular, Dr. A. F. Johnson who read and commented upon the entire manuscript and made many suggestions for improving the clarity of the presentation. Dr. L. W. Morland, Dr. B. E. Read and Dr. R. O. Stafford read individual chapters of the typescript and I am grateful to them for their helpful comments. Finally, my thanks are due to Mrs. Joan Mann and Mrs. Pauline Brice for their care in preparing an excellent typescript.

August, 1972 *F. J. Lockett*

Contents

Preface v

1. Introduction 1

1. Spectrum of viscoelastic materials 2
2. Types of material behaviour 4
3. The physics of solid polymers 10
4. Scope of the present book 13
5. Some basic theory and notation 15

2. Classical Linear Theory of Viscoelasticity 19

1. One-dimensional behaviour 19
2. Interrelations between material functions 26
3. Three-dimensional theory 29
4. Spring-and-dashpot models 33

3. Experimental Determination of Linear Properties 37

1. Direct experimental determination of material functions 37
2. Alternative methods for determining complex modulus 41
3. Experimental considerations 46
4. Typical material data 51
5. Variation of material properties with temperature 55
6. Anisotropic behaviour 57

4. Nonlinear Theory of Viscoelasticity 59

1. Fundamentals of the theory 59
2. Integral constitutive relations 64
3. Differential formulation 68
4. Nonlinear elasticity 69
5. Complex modulus formulation 71

5. One-Dimensional Green–Rivlin Behaviour 74

1. One-dimensional constitutive relation 75
2. Creep and stress-relaxation experiments 79
3. Experimental investigations 83
4. Determination of complex moduli through oscillatory tests 91

6. Three-Dimensional Green–Rivlin Behaviour 95

1. Third-order three-dimensional behaviour 95
2. Incompressible materials: stress-relaxation formulation 98
3. Incompressible materials: creep formulation 105
4. Incompressible plane strain 108
5. Incompressible plane stress 112

7. Single-Integral Constitutive Relations 113

1. BKZ theories 113
2. Finite linear viscoelasticity 121
3. Modified form of Lianis's theory 133
4. Non-isothermal behaviour 137

8. Material Characterization in Special Situations 141

1. Single-step stress-relaxation in general deformation 142
2. Single-step creep under combined load 147
3. Short-time approximation 148
4. Small deformations superposed on finite deformations 151
5. Almost-elastic materials 154

9. Semi-Empirical Approach to Material Characterization 156

1. Empirical representation 156
2. Leaderman's constitutive relation 158
3. Relation between Leaderman's theory and Green–Rivlin formulation 161
4. Product form for material functions 162
5. Other semi-empirical theories 167

6. Range of validity of polynomial theories 170
7. Other multiple-integral theories 173
8. A semi-empirical approach 176
9. Characterization using reduced time 178

References 185
Author Index 189
Subject Index 191

To Anne, Tony and Andrew

Chapter 1 Introduction

Over many years there has been an increasing use of plastics in the manufacture of a wide range of domestic and other articles and as packaging and laminating materials. Plastics components often have had desirable features compared with articles manufactured from traditional materials and have frequently cost less to produce. This latter fact, which arises from ease of manufacture of components by extrusion, injection moulding or similar techniques rather than from the cost of the raw material, has been the main motivator in the increasing use of this type of material. Indeed, although plastics do have a number of desirable characteristics, they have acquired a somewhat dubious reputation as substitute materials which often fail in service. In part this situation has arisen because they have been used in applications where significant mechanical loading has not been envisaged, and have then failed when inadvertently loaded; from a structural point of view they have had less in reserve than traditional materials. In part the situation is due to the peculiar mechanical properties of plastics, in some cases coupled with design procedures which are more relevant to traditional materials. In consequence, plastics have only relatively recently begun to be considered for load bearing applications, although this trend is likely to continue with the wider production and use of reinforced plastics. However, it is evident that satisfactory progress will only be made if adequate methods of material characterization and component design are made available.

Considerable advances have been made in recent years in the development of mathematical theories to model the type of mechanical behaviour exhibited by polymeric materials. It would be a satisfactory state of affairs if these theories could provide the methods of material characterization and component design which are required; unfortunately this is not always the case. When the mechanical behaviour can be represented by a linear infinitesimal-strain theory this can be done, and it is fortunate that this representation is adequate in many practical situations. However, if optimum use of polymeric materials is to be made then situations will arise where characterization of nonlinear behaviour is required. It is unfortunate that, in this case, the relationship between theory and practical application is far from satisfactory. This is partly due to the complexity of nonlinear material behaviour, but, to a large degree, it is due to a lack of communication between theoretical workers and those involved in handling and using materials. It is the purpose of this book to assist in bridging this gap by demonstrating the extent to which present theoretical development can contribute to nonlinear material characterization. In so doing the theoretician should be able to identify better the objective towards which future developments should be directed and the non-mathematician will be provided with an insight into the possibilities and limitations of existing theoretical work.

The present chapter provides an introduction to some basic aspects of the subject. In Section 1 a brief description is given of the spectrum of materials which exhibit viscoelastic characteristics: the most readily observable types of viscoelastic behaviour are discussed in the following section. This is followed by a brief outline of the structural properties of solid polymeric materials and the manner in which this structure is responsible for the observed macroscopic behaviour. Section 4 notes the restrictions which are placed on material behaviour for the purposes of this text and the final section introduces some basic theory and notation. Although these topics are treated in a relatively superficial manner, they provide a common base for discussing a subject which should be considered as an interdisciplinary activity.

1. SPECTRUM OF VISCOELASTIC MATERIALS

The term "viscoelastic" refers, not to a single type of material behaviour, but to a whole spectrum of possible mechanical characteristics. At one extreme are the classical viscous fluids, the most abundant examples of which are air, water and many common fluids. At the other end of the spectrum are the elastic solids; many rubbers and metals exhibit elasticity under normal loading conditions. The study of materials is made more difficult, and more interesting, by the fact that many materials exhibit mechanical behaviour which is a combination of the viscous and elastic extremes. This viscoelastic behaviour may combine viscous and elastic properties in any relative proportions, and thus it forms a continuous spectrum of possible properties passing from one extreme to the other. Near the middle of the spectrum it may be difficult to decide whether a particular material is a solid or a fluid; it may exhibit simultaneously both types of characteristics.

In order to give some idea of the types of material which are viscoelastic, it is useful to discuss three examples which span the spectrum of viscoelastic behaviour. Near the elastic extreme are the solid polymers such as nylon, rigid polystyrene, or any of the materials commonly referred to as plastics. These viscoelastic solids may appear, upon a casual inspection, to behave elastically; however, a more careful examination reveals that they are also capable of flowing. Thus their mechanical behaviour is essentially elastic but contains some viscous characteristics, and they occupy a position on the spectrum near the elastic extreme. Some metals and filled rubbers occupy a similar position on the spectrum. An example of a viscoelastic fluid is provided by a polymer solution in which a small amount of solid polymer is dissolved in a purely viscous fluid. The solid polymer brings to the solution some of its elastic characteristics, and the resulting fluid occupies a position near the viscous extreme. Some of the viscoelastic characteristics of these fluids will be discussed in the next section. Finally a material is considered which occupies a more-or-less central position on the spectrum. When a polymer is brought into a molten state by a rise in temperature its ability to flow is increased and it is obvious that a suitable temperature rise will produce a material in which solid and fluid-like properties are equally important. This example of viscoelastic behaviour is particularly important in the forming process of many plastics components where extrusion or injection moulding of a molten polymer is involved.

Before concluding this section it may be necessary to correct a possible misconception arising from the previous discussion. The examples quoted above refer to materials under *normal* laboratory or practical situations: this requires some amplification. In fact the position on the spectrum is not determined by the material alone, but also by the operating conditions. Thus a plastics material may occupy a central location or a position near the elastic extreme under different temperature conditions. However, as will now be shown, a material may exhibit quite different types of behaviour merely under different mechanical conditions.

Many metals behave elastically under normal conditions. Thus steel used in the construction of a bridge will behave elastically under the loading situations to which the structure is likely to be subjected. However, if that same material were used as armour plate, and this were struck by a very-high-velocity projectile, the situation would be quite different. In this case the mode of deformation of the steel would be more akin to that produced in water when a stone is dropped into a pond. The material in the immediate neighbourhood of the impact would behave more like a viscous fluid than an elastic solid, and it can be seen that this material can occupy very different positions on the spectrum depending on the conditions to which it is subjected.

A second example, which offers a useful comparison, is the material sold in toy and novelty shops as "silly putty". When this material is deformed at slow rates it behaves like normal putty, that is, as a highly viscous fluid. However, a suitably shaped piece of this material will bounce like an elastic ball. Thus, in the first example, the material behaviour is essentially elastic at the low rates of loading and viscous at the higher rates; in the second example the converse is true. This contrast is readily explained at a microscopic (i.e. atomic or molecular) level, but the essential point here is that it illustrates that a knowledge of both the material and its intended application is required in order to determine which part of the viscoelastic spectrum is involved. For this reason a discussion of viscoelastic characteristics should be based on various types of viscoelastic *behaviour* rather than on different types of viscoelastic *materials*. It will remain for experience, physical intuition and experimental investigation to determine which type of behaviour a particular material will exhibit under a given set of conditions. The more readily observable types of mechanical behaviour which may be exhibited by viscoelastic materials will now be described in further detail.

2. TYPES OF MATERIAL BEHAVIOUR

A. Creep and Recovery

It was noted in the previous section that viscoelastic solids such as plastics may exhibit flow in addition to their elastic characteristics. This flow behaviour is called "creep" and it may be readily observed in a simple experiment. Consider a rod of material which is loaded in tension by a time-dependent force $L(t)$ and suppose that the resulting elongation $d(t)$ is measured. In particular let $L(t)$ be zero up to some time, which is defined as $t = 0$, and then assume a constant value L for $t \geqslant 0$. If the material is elastic the elongation $d(t)$ will have a similar time variation provided that inertia effects are negligible; there will be no elongation until the load is applied at $t = 0$, and thereafter the constant load L will produce some constant elongation in the rod which depends on the stiffness of the material. In practice inertia effects would cause some deviation from this response immediately after the sudden load application; for the purpose of this discussion it will be assumed that some mechanism is provided to damp-out these deviations.

When the material is viscoelastic the elongation $d(t)$ has a form similar to that shown schematically as curve ABC in Fig. 1. Again, there is no elongation prior to $t = 0$, and upon application of the load there is an instantaneous elastic response corresponding to OA. However, whereas the load L is maintained at a constant value for $t \geqslant 0$, the elongation $d(t)$ steadily increases; this is the phenomenon of creep under constant load. For some materials the curve ABC will eventually approach a zero slope so that $d(t)$ reaches an equilibrium or long-time constant value. For other materials the curve may approach a constant non-zero slope after which $d(t)$ increases indefinitely but in a linearly time-dependent manner. This latter type of behaviour is precisely the response (for all time $t \geqslant 0$) of a purely viscous fluid and, for this reason, materials with equilibrium values of $d(t)$ may be referred to as solids and those which exhibit continued deformation can be called fluids. However, this definition is not entirely satisfactory in practice since it may not be possible to decide in the time scale of an experiment whether the curve ABC tends to a zero slope or to a small non-zero value.

If the load L is removed at some time $t = T$ an elastic material will immediately return to its original undeformed state ($d = 0$), subject to neglect of inertia effects. A viscoelastic material will follow a delayed recovery path similar to that shown schematically as BDE in Fig. 1. This is made up of an instantaneous elastic component BD followed by a gradual recovery DE. If the material is a solid in the sense described above, the curve DE will tend to zero, whereas a fluid will exhibit some residual deformation as $t \to \infty$. Again this is an unsatisfactory means of defining solids and fluids since, in the time-scale of an experiment, it may not be possible to say definitely whether there is a small or zero residual.

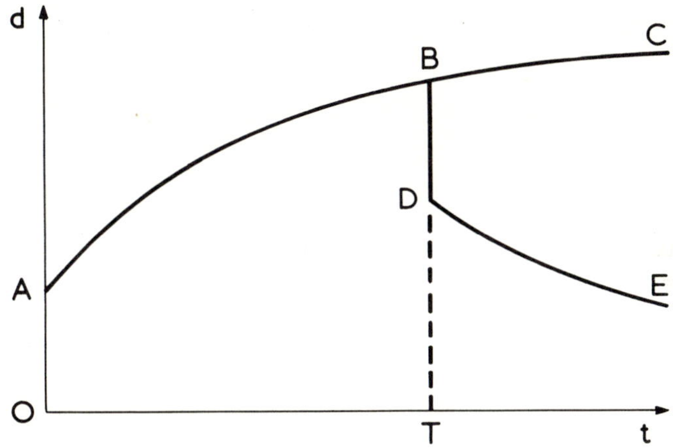

Fig. 1. Creep and recovery.

B. Stress-Relaxation

This type of material behaviour is the dual of creep behaviour although it is possible for a material to exhibit one of these characteristics but only a degenerate form of the other. Stress-relaxation may be demonstrated for a rod in tension by *applying* a particular *elongation* $d(t)$ and measuring the load $L(t)$ required to maintain this deformation. If d is zero for $t < 0$ and constant for $t \geqslant 0$, then an elastic material would require a load $L(t)$ of the same time-dependent form. A viscoelastic material requires some load (OA in Fig. 2) to produce the deformation at $t = 0$, but the load required to maintain this deformation decreases with time as shown by ABC in Fig. 2; this is the phenomenon known as stress-relaxation. In principle, a viscoelastic solid would require a non-zero equilibrium load, but a fluid would eventually reach a stage where it could remain in the deformed state without a load L. The extreme case is a purely viscous fluid which requires an impulsive load to produce the deformation (in practice it would take a finite time to achieve this) but no load is required once that deformation is obtained.

In order to remove the deformation at some time $t = T$ it is sufficient for an elastic material to remove the load at that instant. However, for a viscoelastic material it will be necessary to apply the load DE shown in Fig. 2. This load is compressive if the load applied in the interval $0 \leqslant t < T$ is tensile, but its magnitude eventually relaxes to zero as the material "forgets" the deformation to which it had been subjected.

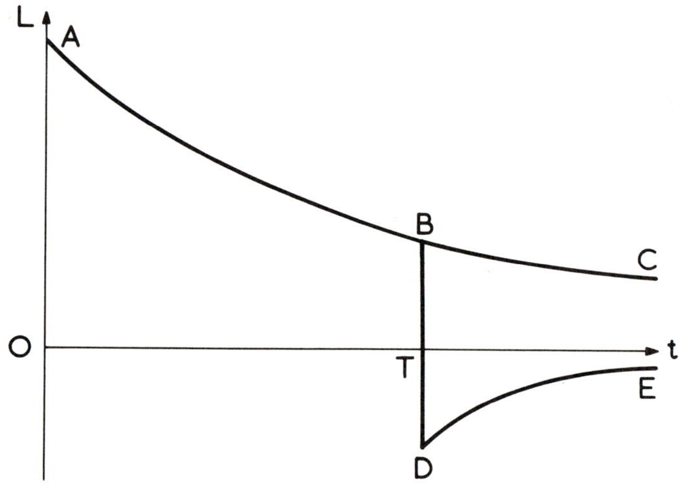

Fig. 2. Stress-relaxation.

C. Rate-Dependent Viscosity

Consider a material contained between two parallel plates, one of which is stationary whereas the other moves with constant velocity V in its own plane. When the motion is steady and homogeneous the ratio V/h, where h is the distance between the plates, is the velocity gradient in the material and is equal, in this simple situation, to the shear strain-rate. In order to maintain the motion it is necessary to apply some load F to the plates in the direction of relative motion and, if the material is a Newtonian viscous fluid, the load F will be proportional to the shear strain-rate V/h. Apart from a purely geometrical parameter, the constant of proportionality is the viscosity of the fluid. For more complicated types of fluid behaviour the ratio of load to strain-rate is not constant and the fluid is said to exhibit rate-dependent viscosity. In particular, if this ratio decreases with increasing strain-rate the fluid is said to be shear-thinning or pseudoplastic. If the ratio increases with increasing strain-rate the fluid is shear-thickening or dilatant. It should be noted that in these definitions the ratio of load to strain-rate is assumed to depend only on the current strain-rate. Thus, if a fluid is made to flow first at one rate, then at a second rate, and finally at the first rate again, the apparent viscosity is assumed to be the same in the first and last cases. This is a different phenomenon from that exhibited by non-drip paints, where shearing of the fluid by stirring permanently reduces the viscosity at all rates. This behaviour is called thixotropy and in this case is due to mechanical degradation of the fluid. A second example of thixotropy, involving a permanent increase in viscosity, is provided by cream which thickens when subjected to the shearing forces of whipping. The reader will find a fuller discussion of these and similar concepts in the book by Severs (1962).

D. Normal Stresses

In the simple shear-flow experiment referred to above the forces F are the only forces required to support the motion when the material is a Newtonian viscous fluid (gravitational body forces are assumed to be negligible and the case is considered where no lateral forces are applied to the fluid). For more general fluid behaviour the two plates will tend to move closer or further apart unless forces are applied normal to the plates. In more general terms, the state of stress in a fluid undergoing shear may be different from a shear stress plus a pressure acting equally in all directions. Nonlinear viscous fluids exhibit some normal stress differences, but they are of a less general form than the visco-elastic normal stresses mentioned here.

A simple demonstration of the effects of normal stresses is obtained by containing a suitable fluid in the annular region between two concentric cylinders (see Fig. 3). Consider first a situation which it would not be easy to achieve in practice. When the inner cylinder is rotated about its axis, and a surface pressure distribution is applied to maintain a plane fluid surface, it is readily shown that a velocity distribution is set up in the fluid which corresponds to shear rates which decrease with increasing radial position. The contribution to the surface pressure distribution which arises from normal stress effects is then a decreasing function of radial position; for a Newtonian viscous fluid this contribution vanishes. The other contribution arises from the effects of centrifugal forces and is an increasing function of radial position. In suitable circumstances the total pressure distribution has a minimum at some location between the cylinders, and increases as both cylinders are approached. In practice it would be difficult to apply this required surface pressure distribution, but it is a simple matter to allow it to take the uniform atmospheric value. It is intuitively reasonable that, when this change of pressure distribution takes place, the fluid surface should take up a form like that shown in Fig. 3; the fluid tends to climb the outer cylinder due to centrifugal effects and the inner cylinder due to normal

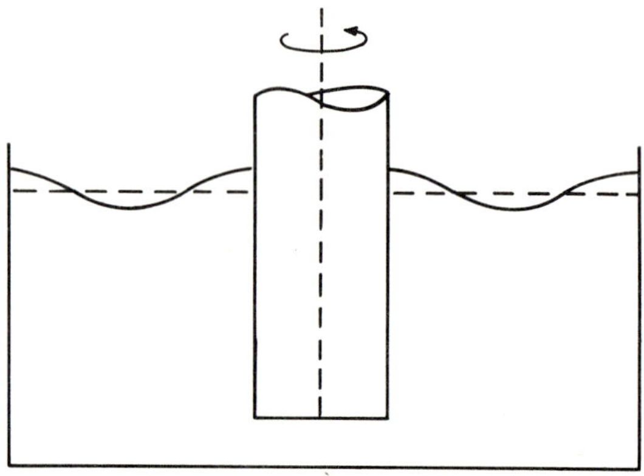

Fig. 3. Experimental demonstration of the Weissenberg effect.

stress effects. This latter phenomenon, which is called the Weissenberg effect, is readily observable (see, for example, Lodge, 1964, p. 232).

E. Elastic Recoil

When a Newtonian fluid is set in motion and then brought to rest in a new configuration it will show no tendency to return to its original configuration provided that no external forces are applied. However, a viscoelastic fluid may exhibit some elastic recovery or recoil. Thus, if a viscoelastic fluid is contained in the annulus between two concentric cylinders which are given a relative angular displacement, then the cylinders will tend to move so as to reduce their angular displacement when the loads are removed.

F. Distinction Between Solids and Fluids

The reader will have noted that the word "material" has been replaced by either "solid" or "fluid" in the description of the various types of material behaviour, thereby implying that the particular characteristics are observed only, or more readily, in one or other type of material. However, it has been noted that under one definition of the distinction between solids and fluids it may be difficult to decide, in the time scale of an experiment, whether a particular material is a solid or a fluid. Since a possible confusion in terminology exists, and since this text is concerned primarily with viscoelastic solids, it is essential that the distinction should be more firmly defined.

The purpose of this text is to discuss the behaviour of materials which may be used in load-bearing components of structures. It is a prerequisite for such materials that they should be able to bear their own weight and, in particular, it should be possible to fabricate a test specimen which does not need to be supported in some form of container. It is in this sense that the present text is confined to solid materials, but these materials may still be fluids in the sense discussed in Sub-sections 2A and 2B and in a sense to be discussed more precisely in Chapter 4.

The main impacts of restricting attention to solid materials are to place restrictions on the types of experimental methods considered and on the parameters used to characterize material behaviour. Thus extension of a rod of material, as discussed in Sub-sections 2A and 2B, is not appropriate to a fluid without some modification; a similar discussion could be applied to shear of a fluid between concentric cylinders where L is replaced by the applied torque and d is replaced by the relative angular displacement. Similarly the latter experimental configuration does not lend itself readily for use with solids. For solid materials it will usually be more appropriate to use material parameters which generalize those of elastic solids, rather than to use parameters such as viscosity which are more appropriate to fluids.

The types of material and material behaviour which are excluded from the present text involve rate-dependent viscosity, normal stress effects, and other fluid characteristics. These topics are dealt with very adequately, from both a theoretical and experimental point of view, in the books by Lodge (1964), Pearson (1966) and Coleman *et al.* (1966).

G. Yield and Plasticity

Many materials exhibit a yield point (usually characterized by a critical stress value) beyond which the mechanical behaviour is radically different from that involved below yield. Thus many metals behave elastically below yield and plastically after yield; this discontinuous type of material behaviour will not be treated in this book. However, a few words of explanation may be necessary to justify the exclusion of *plasticity* from a discussion of material behaviour which is highly appropriate to materials known as *plastics.*

Plastics take their name from their ability to undergo plastic deformation when subjected to sufficiently high temperatures and pressures. This property allows the material to be deformed by some kind of shaping or moulding process to produce an end-product which retains its shape; this is a main practical attraction of plastics since the required temperatures and pressures are not difficult to achieve. However, when an article or component is used in service it is almost invariably desirable that plastic deformation should not occur, whether or not plastic deformation was involved during manufacture. Thus the behaviour of materials and components in service does not involve consideration of plasticity; but the properties of the material *in* the component may depend on the manufacturing process.

3. THE PHYSICS OF SOLID POLYMERS

The type of mechanical behaviour considered in this text is most commonly exhibited by plastics and some rubbers; that is, by materials having a polymeric structure. In the engineering approach adopted here attempts are made to model the macroscopic behaviour of the materials and this requires them to be treated as continua. Although macroscopic properties undoubtedly depend on the microscopic structure of the material, no knowledge of this structure is required in the continuum approach. However, it is of interest to give a brief description of the structural properties which give rise to the observed mechanical behaviour. Further, it will be apparent from later chapters that the continuum approach has not been wholly successful in treating nonlinear mechanical behaviour. It is evident that microscopic studies could contribute to continuum theories, for example, by indicating the physical significance of certain parameters, but this will not be attempted here.

A. Molecular Chain Structure

Polymeric materials are distinguished by the long chain-like nature of their molecules. The synthesis of these chain molecules is effected by polymerization reactions of certain small molecules known as monomers. The polymerization reactions may involve the repeated addition or condensation of the monomer, and the resulting repeat unit of the chain is thus similar to, though not identical with, the original monomer. Most industrial plastics have molecules based on a carbon backbone, each carbon atom being linked to the two adjacent carbon atoms by covalent bonds. Different plastics are formed from monomers containing different substituent groups, and the resulting polymer chains thus differ in the nature of the substituents attached to the carbon atoms within each repeat unit. Typical examples of plastics containing carbon backbone chains are polyethylene, $[—CH_2—CH_2—]_n$, and polyvinyl chloride, $[—CH_2—CHCl—]_n$, where n is the so-called degree of polymerization. Not all polymers are based on a carbon backbone; the silicone rubbers, $[—Si(CH_3)_2—O—]_n$, which possess mechanical properties very similar to those of carbon-based rubbers, have polymeric chains comprising alternate silicon and oxygen atoms within the backbone. Most other polymeric structures based on the —Si—O— repeat unit, such as glass, have properties which differ considerably from the common synthetic carbon-based polymers under similar circumstances. These differences generally result from the presence of ionic substituents in the case of linear chains, or from the existence of two or three dimensional polymeric structures, rather than from any marked differences between the properties of the carbon and silicon–oxygen backbones. Both classes of material exhibit viscoelastic behaviour, however, and are therefore of interest to the subject of this text.

In addition to the homopolymers which are formed from identical repeat units, copolymers can be produced by polymerizing together two or more suitable monomeric components. When the different repeat units are arranged randomly within the chain, the polymer is termed a random copolymer, and tends to exhibit properties intermediate between the homopolymers formed from the respective repeat units. If the chains comprise groups or blocks of the different repeat units, the resulting material is referred to as a block copolymer and, since the blocks frequently tend to be incompatible, multiphase materials are often produced which have properties resembling those of a mixture of the different homopolymers.

The long chain-like molecules of a polymer may be linear in the sense that they form long threads, they may be branched, or they may be crosslinked in which molecular chains are joined at various points along their length to other chains. In the latter case the whole body of material may constitute a giant molecular network. A few words of qualification should be added to the use of the word long in relation to the molecular chains. Although a typical polymer chain may have a degree of polymerization of tens of thousands, its total length if pulled out straight would be only of the order of one hundredth of a millimetre. Since molecular chains never occur in this extended form, but usually adopt a tangled ball-like configuration, their physical dimensions are several orders of magnitude smaller than this.

Most polymers are capable of existing in either an amorphous or partly crystalline form; indeed, most material samples will have both amorphous and crystalline regions. In the amorphous state the molecular chains are arranged in a purely random fashion, forming a tangled mass. When the chain structure is extremely symmetrical, or when the forces exerted between chains are very strong, it is possible for the material to adopt a more systematic configuration; that is, a crystalline form.

B. Thermoplastic and Thermosetting Materials

The distinction between thermoplastic and thermosetting polymers is essentially governed by the un-crosslinked nature of the former, which allows relative motion of the molecular chains when the material is under load, and the crosslinked nature of the latter which inhibits such motion. The prefix thermo- is used in the two terms, since some thermosetting materials may be un-crosslinked initially, the crosslinks being formed chemically upon the first application of significant heating. The material will flow under pressure in the early stages of this heating, and can be used to form an article or component, but on cooling or reheating it behaves like, and is, a crosslinked material.

Thermoplastic materials do not form crosslinks and may be caused to flow repeatedly by application of pressure, with or without temperature rise.

It should not be inferred from the discussion above that crosslinked materials are incapable of flow; but they are incapable of indefinite flow. To see this it is necessary to consider briefly the mechanisms by which polymeric materials are able to deform. It has been noted that they consist, in the amorphous form, of long thread-like molecules forming a tangled mass. The material is able to deform through two mechanisms: individual chains may extend and there may be relative motion between adjacent chains. The former is essentially an elastic mechanism whereas the latter is essentially viscous in nature, however it is evident that each mechanism will tend to induce the other as neighbouring chains interact. Both crosslinked and un-crosslinked materials may deform under both mechanisms but, in the case of the former, the presence of crosslinks will eventually inhibit further deformation. In the case of a completely cured thermosetting material the flow mechanism may be negligible, but it can be significant for a material with a relatively low degree of crosslinking, such as a vulcanized rubber. In a thermoplastic material the initial stages of deformation involve both mechanisms and are truly viscoelastic: when the chains are relatively well extended the viscous mechanism becomes most dominant, and the behaviour is best described as *plastic* flow.

Since it is the means by which rubbers deform, the mechanism of chain uncoiling is referred to as rubberlike elasticity. Large recoverable deformations can arise through this mechanism; in elastomers such as natural rubber recoverable deformations of several hundred per cent can occur, and plastics may exhibit strains of several per cent before non-recoverable deformation occurs. This behaviour is in sharp contrast to that of metals where elastic deformations arise from distortion of atomic bonds and elastic strains in excess of approximately 0.2% are not possible.

When a small amount of solvent is added to a polymer, plasticization occurs and the solvent is described as a plasticizer. The additions of plasticizer act as a lubricant which allows chains to slip under stress; this helps to produce a tougher material since energy is absorbed readily by the more mobile molecules. A material formed of a highly tangled mass of molecular coils, without the addition of a plasticizer, tends to be brittle due to the difficulty of energy absorption by molecular rearrangement.

C. Glass Transition

A notable feature of the mechanical properties of polymers is the considerable variation of these properties with variation in temperature, the most pronounced changes occurring at relatively modest temperatures compared with the melting temperatures of metals. At all temperatures the configurations of the molecules are not stationary, but execute a kind of Brownian motion. Below a certain temperature this motion is relatively weak and is incapable of overcoming the inter-molecular hindrances caused by the presence of neighbouring molecules. At these temperatures the material is relatively hard and unyielding, and it is said to be in the glasslike state. In the neighbourhood of the glass-transition temperature the thermal energy of motion becomes sufficient to overcome the barriers to molecular rotation and movement, and softening of the material occurs. As temperature is increased this softening does not occur as suddenly as the melting of a metal, but develops gradually over a temperature range of perhaps 20°C. At temperatures above the glass transition region chain rotation and uncoiling can take place with relatively little viscous effect, so that rubberlike behaviour is observed. In the rubbery region the stiffness of the material is considerably less than in the glassy region; the values of stiffness may differ by several orders of magnitude.

Polymeric materials exhibit viscoelastic behaviour in the glassy and rubbery regions, but this behaviour is most pronounced in the transition region. At normal ambient temperatures plastics are in the glassy state, whereas elastomeric materials are in the rubbery state. Plastics can be brought into the rubbery state by heating to beyond the glass transition temperature; for many plastics this is less than the temperature of boiling water, and few plastics have glass transition temperatures greater than 200°C. In a similar manner elastomers become hard glassy materials with a modest decrease in temperature.

4. SCOPE OF THE PRESENT BOOK

This book is concerned with the gross mechanical properties of materials and their relevance to the design of components including prediction of component behaviour in service. For a given material this requires the abilities to model the mechanical behaviour through a mathematical or engineering theory, to design experimental tests for determining material parameters, and to develop techniques for solving the boundary value problems associated with particular applications. It is obvious that the first two must precede the latter; whether theory should precede experimentation or vice versa depends on one's point of view; a main conclusion of this text is that these two approaches should be very much interrelated for the complex material behaviour discussed here. The main objective should be to determine a model which is as simple as possible consistent with the ability to represent experimental data to an acceptable degree of accuracy. Techniques for solving boundary value problems, and solution of significant problems, will not be discussed here; the linear case is covered admirably in other texts (for example, Christensen, 1971) whereas the nonlinear case is not at a sufficient state of development.

Empirical theories, which have no rational scientific basis, but are invented to model particular sets of experimental data, will not be discussed. Thus theories will be considered which express a given set of physical axioms in a quantitative manner. However, any such theory can only be as good as the physical axioms upon which it is based, and these may be insufficiently precise. Thus the validity of a particular theory should be judged as much by its ability to model real behaviour as by its mathematical rigour. In addition it is the author's view, in contrast to that of Truesdell (1965), that it is as much the duty of the theoretician to assist in simplifying and applying these theories as it is the duty of the experimentalist or designer who wishes to use them.

It is tempting to seek a material model which can be used in a wide range of conditions of loading, temperature, etc. and many contributions to the literature have proposed theories including relativistic, electromagnetic and other effects. In some cases linear theories of this type may be sufficiently simple to apply. However, when nonlinear theories are considered, it soon becomes obvious that nonlinearity of mechanical properties is, in itself, at the limits of present understanding. For this reason the present book is concerned primarily with isothermal isotropic behaviour of a purely mechanical type.

It has already been remarked that, within the range of solid behaviour, materials such as plastics have mechanical properties which are extremely temperature-sensitive in the neighbourhood of the glass transition. Since this transition occurs at modest temperatures, it is obvious that thermal effects will be significant in many applications. If the mechanical behaviour is of a type known as thermo-rheologically simple (see Chapter 3) some analyses of practical situations can be performed (see Christensen, 1971). However, relatively little progress has been made in nonlinear thermo-mechanical characterization and analysis. Of course, thermal effects are usually most pronounced during polymer melt processing, but this has already been excluded since plastic flow is not considered here.

Anisotropy of mechanical properties is undoubtedly of real practical importance both from the point of view of the orientation produced during processing, for example in fibre drawing, and in view of the practical importance of materials such as glass-fibre reinforced plastics. The linear case will be discussed briefly in Chapters 2 and 3, but it will again be necessary, through lack of development, to postpone the nonlinear anisotropic case to a later edition of this book.

Finally, it is relevant to reiterate some restrictions made earlier. In general the discussion is concerned with solid materials, although it has been noted that fluid-motivated theories may nevertheless be appropriate to describe material behaviour. It has also been stated that behaviour during processing is excluded whenever this involves plastic flow or thermal effects. Material properties before and after processing will generally be different and, since they cannot be related by present theory, it is the mechanical properties of the processed material which are relevant to a design calculation or performance prediction. It should be noted that failure is not included in the discussion; this subject is dealt with by Andrews (1968).

5. SOME BASIC THEORY AND NOTATION

In this section some basic equations are recalled which are valid for a wide range of material behaviour including that of the viscoelastic solids considered in this text. The constitutive equations, which complete the field equations for isothermal mechanical behaviour, are recalled for linear elastic solids and Newtonian fluids. The remainder of the book is concerned with discussing the corresponding constitutive equations for viscoelastic materials. To facilitate the discussion a fixed system of rectangular Cartesian coordinates will be used throughout; curvilinear systems introduce further complexities into the equations but no new physical features. Since this book is not concerned with the solution of boundary value problems, nothing is lost by restricting attention to the simpler system.

Consider a rectangular Cartesian system (x_1, x_2, x_3) in which the coordinates of a typical material point are denoted by (X_1, X_2, X_3) in some convenient reference configuration. During deformation of the body of material the typical material point will occupy new locations; the notation $x_i(X_j, t)$ is used to denote the location at time t of the particle which was located at coordinates X_j $(j = 1, 2, 3)$ in the reference configuration. The components of the displacement vector $\mathbf{u} = (u_1, u_2, u_3)$ are observed to be

$$u_i = x_i - X_i \tag{1.1}$$

and the components of the velocity vector $\mathbf{v} = (v_1, v_2, v_3)$ are

$$v_i = Du_i = \left.\frac{\partial u_i}{\partial t}\right|_{X_j = \text{constant}} \tag{1.2}$$

where D denotes the material time derivative.

The concept of stress must now be introduced and, since applications will be concerned with the equilibrium or motion of the material in its *deformed* state, stress is defined with respect to the deformed configuration. Consider an imaginary plane surface in the deformed body of material with surface normal $\mathbf{n}$ passing through a typical material point. The part of the material into which $\mathbf{n}$ is directed exerts a force on the material on the other side of the surface. The vector $\mathbf{T}$ is used to denote the limit at the typical point of the resultant force per unit area as the area considered tends to zero. The value of $\mathbf{T}$ depends on the orientation of the plane surface, that is on **n**; and it can be shown that the transformation connecting **T** and **n** is expressible in the form

$$T_i = \sigma_{ij} n_j \tag{1.3}$$

where summation is implied over all values of the repeated suffix j. The matrix of coefficients σ_{ij} in this linear transformation are the components of the stress tensor.

The stress components are related to the displacement or velocity components through the equations of motion which express the balance of linear and angular momentum

$$\sigma_{ij,j} + F_i = \rho D^2 u_i = \rho D v_i$$

$$\sigma_{ij} = \sigma_{ji}. \tag{1.4}$$

Here the suffix $,j$ denotes the derivative $\partial/\partial x_j$, which arises because balance of momentum is applied to the deformed material, ρ is the material density and F_i denote the components of the applied body force defined per unit volume in the deformed body. For a compressible material ρ can be related to the velocity field, or to the displacement field and density ρ_0 in the reference configuration, through the mass conservation relations

$$\frac{\partial \rho}{\partial t} + (\rho v_j)_{,j} = 0 \tag{1.5}$$

$$\rho_0 = \rho \det \left| \frac{\partial x_i}{\partial X_k} \right|. \tag{1.6}$$

When the material can be approximated adequately by the assumption of incompressibility the density ρ is a constant parameter and equations (1.5) and (1.6) impose constraints on the velocity or displacement fields.

It will be noted that straightforward evaluation of the first term in (1.4) requires the stress components to be expressed as functions of x_i and t, whereas evaluation of the inertia terms through the defintion of D contained in (1.2) requires u_i or v_i to be expressed as functions of X_i and t. Since it is usually convenient to express all dependent variables in terms of the same independent variables, it is necessary to make use of one of the relations

$$\frac{\partial}{\partial x_j} = \frac{\partial}{\partial X_k}\frac{\partial X_k}{\partial x_j}, \qquad D = \frac{\partial}{\partial t} + v_k \frac{\partial}{\partial x_k}. \tag{1.7}$$

It will be seen later than this distinction does not arise in the classical linear infinitesimal theories. For equilibrium conditions the inertia terms on the right-hand side of (1.4) vanish and a conflict of independent variables does not arise. However, it is usually most convenient to use the reference coordinates X_i as independent variables when considering solid behaviour, and the first of (1.7) is then required in (1.4).

In general equations (1.4), together with (1.5) or (1.6), provide only four equations connecting six stress components and the density with three displacement or velocity components. The field equations are completed by the addition of constitutive relations which characterize the particular material behaviour being considered. The form of these equations for viscoelastic solids, and the associated experimental requirements, is the subject of the remainder of this text. As a preliminary the equations describing the extreme cases of linear elastic and Newtonian viscous behaviour will now be recalled. However, it should be borne in mind that all of the equations presented up to this point are valid for a wide range of material behaviour, and it is the constitutive relations which restrict attention to a particular type of material.

In the classical theory of infinitesimal linear elasticity the further relations connecting stress and displacement are provided through the intermediary of a strain tensor ϵ_{ij} which is defined as

$$\epsilon_{ij} = \tfrac{1}{2}(u_{i,j} + u_{j,i}). \qquad (1.8)$$

In the linearized theory involving small displacements, no distinction is made between the independent variables X_i and x_i; it is usual to replace x_i by X_i, that is, to define all quantities and derivatives with respect to the reference configuration. The constitutive relation for a linear elastic material can now be written in the form

$$\sigma_{ij} = c_{ijkl}\,\epsilon_{kl} \qquad (1.9)$$

where summation is implied over both of the repeated suffixes k and l. The coefficients c_{ijkl} are the elastic moduli of the material; due to symmetry of the stress and strain tensors and due to thermodynamic arguments it can be shown that

$$c_{ijkl} = c_{jikl} = c_{ijlk} = c_{klij} \qquad (1.10)$$

and it follows that the modulus tensor has only 21 independent components. In its general form (1.9) models anisotropic mechanical behaviour; various possible forms of linear anisotropy are discussed in texts on that subject (see, for example, Lekhnitskii, 1963). Here attention is restricted to the special form which (1.9) takes when the material is isotropic:

$$\sigma_{ij} = \lambda \Delta \delta_{ij} + 2\mu\epsilon_{ij}, \qquad \Delta = \epsilon_{kk} \qquad (1.11)$$

which can also be written in the convenient matrix form

$$\boldsymbol{\sigma} = \lambda\Delta\mathbf{I} + 2\mu\boldsymbol{\epsilon}, \qquad \Delta = \operatorname{tr}\boldsymbol{\epsilon} \qquad (1.12)$$

where $\mathbf{f} = \|f_{ij}\|$, $\mathbf{I}$ is the unit matrix and tr denotes the trace of a matrix. The material parameters λ and μ are known as Lamé's constants; μ is the shear modulus of the material and is sometimes denoted by the symbol G. Other alternative parameters which may be used are Young's modulus E, Poisson's ratio ν and bulk modulus K: they are defined in terms of λ and μ through

$$E = \mu(3\lambda + 2\mu)/(\lambda + \mu), \qquad \nu = \lambda/2(\lambda + \mu), \qquad K = \lambda + \tfrac{2}{3}\mu. \tag{1.13}$$

When a material is assumed to be incompressible it is possible first to assume compressibility and then allow K (or alternatively λ) to tend to infinity in the *solution* of a boundary value problem. However, it is obvious that this limit cannot simply be taken in equation (1.12) before the solution is obtained. To avoid having to take the limit of the solution for a compressible material, it may be noted that a deformed incompressible material is unaffected by an additional hydrostatic pressure. This implies that, for an incompressible material, the constitutive relation can only define $\boldsymbol{\sigma}$ in terms of $\boldsymbol{\epsilon}$ to within an arbitrary hydrostatic pressure and (1.12) should be replaced by

$$\boldsymbol{\sigma} = -p\mathbf{I} + 2\mu\boldsymbol{\epsilon} \tag{1.14}$$

where the pressure p is not a material parameter but must be determined from the solution of the field equations with appropriate boundary conditions.

In the classical theory of Newtonian viscous fluids the stress components are related to the velocity components through a strain *rate* tensor d_{ij} defined by

$$d_{ij} = \tfrac{1}{2}(v_{i,j} + v_{j,i}) \tag{1.15}$$

and, in this case, the independent variables are chosen to be the x_i. Constitutive relations connecting $\boldsymbol{\sigma}$ and $\mathbf{d}$ can be written down and, in the simpler case of an incompressible isotropic fluid, they take the form

$$\boldsymbol{\sigma} = -p\mathbf{I} + 2\eta\mathbf{d} \tag{1.16}$$

where η is the fluid viscosity. In some treatments of this subject it is usual to omit the factors $\frac{1}{2}$ and 2 from equations (1.15) and (1.16).

For fuller discussions of these field equations and their solution the reader is referred to the standard texts on elasticity (for example, Sokolnikoff, 1956) and fluid dynamics (for example, Landau and Lifshitz, 1959). The remainder of the present text is concerned with the equations which replace (1.12), (1.14) and (1.16) when the material behaviour is that of a viscoelastic solid, and with the experimental methods which are available or under development for determining the material parameters which replace λ, μ, E, η, etc. in these theories.

Chapter 2 *Classical Linear Theory of Viscoelasticity*

When the mechanical behaviour of a material is linear, in the sense that twice the load produces twice the response, and when strains are small so that classical definitions of stress and strain are adequate, the mathematical theory of viscoelasticity is complete and it can be applied satisfactorily to model real materials and solve problems of practical interest. In this chapter some of the main results of the theory are summarized, and in the next a description is given of the experimental methods that can be used with the theory to characterize linear mechanical behaviour. The first two sections of the present chapter are concerned with purely one-dimensional behaviour through which the essential features of the theory can be introduced. Extension to three-dimensional behaviour is given in Section 3 and spring-and-dashpot models are discussed in Section 4.

1. ONE-DIMENSIONAL BEHAVIOUR

A. General Discussion

Consider a one-dimensional situation in which the stress $\sigma(X,t)$ and strain $\epsilon(X,t)$ depend on a single material coordinate X and on time t. At this stage the precise nature of the one-dimensional system need not be specified; it may represent extension of a fibre, pure shear, bulk deformation, or some other system in which the relation between a single stress measure and a single strain measure is of interest.

When an elastic material is deformed, experiment has shown that the stress in the material at any time is proportional to the strain at the same location and time, and does not depend on the past history of strains to which the material has been subjected. This enables a linear theory of elasticity to be formulated upon the assumption that stress $\sigma(X,t)$ is a function of $\epsilon(X,t)$. Similarly, experimental experience suggests that the stress in a linear viscous fluid is dependent upon the current rate at which strain is changing with time, that is $\sigma(X,t)$ can be expressed as a function of $\partial\epsilon(X,t)/\partial t$. Thus a viscous fluid "remembers" the past history of deformation only to an extent which can be expressed in terms of the current strain rate. The two simple physical observations mentioned above are the basis upon which the classical mathematical theories of elastic solids and viscous fluids are built.

Now practical experience with materials which exhibit behaviour which the physicist or engineer would call viscoelastic, suggests that these materials remember previous deformations to which they have been subjected in a more general manner. Thus it can be postulated that linear viscoelastic materials are such that the stress $\sigma(X,t)$ depends on the strains $\epsilon(X,\tau)$ to which the material was subjected at all times τ up to and including the time t. Mathematically this can be expressed in the form

$$\sigma(X,t) = \mathop{\mathscr{F}}_{\tau=-\infty}^{t} \{\epsilon(X,\tau)\} \tag{2.1}$$

where $\mathscr{F}$ is some functional which is to be determined by experiment. Two points are worth emphasizing in connection with (2.1). Since the functional $\mathscr{F}$ does not have the spatial variable X as an explicit argument, identical strain histories at any two points in the same material will produce identical stresses at these two points. Thus (2.1) refers to a homogeneous body of material. In

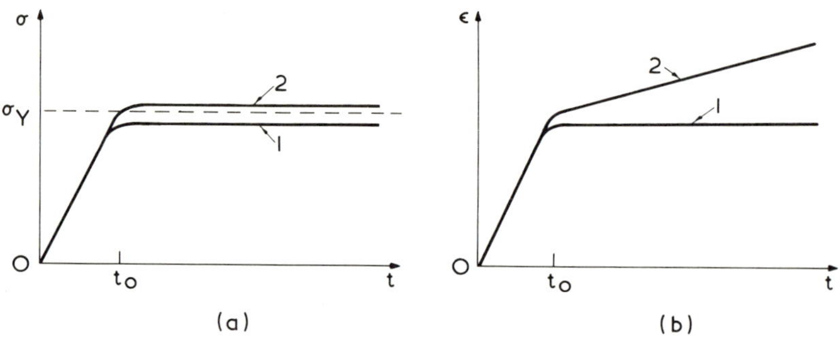

Fig. 1. Strain responses (b) of an elastic-plastic material to two loading histories (a); σ_Y is the yield stress.

addition, it is noted that the stress at some point X has been expressed in terms of the history of strain *at that point*. Obviously the state of stress and strain in a body is affected by that in neighbouring parts of the body, but the present theory assumes that both stress and strain change in such a way as to preserve the relationship between stress and strain at a point. Having noted the implied restrictions on the theory, dependence on spatial variation will be suppressed in future equations.

Equation (2.1) is the starting point for the mathematical theory of one-dimensional viscoelasticity. However it should be noted that this theory will model real viscoelastic materials only if the physical postulate preceding (2.1) is an adequate representation of the observed behaviour. Since the objective should be to model the behaviour of real materials, whose characteristics have been given the name viscoelastic, it is necessary that theories derived from (2.1) should be tested by experiment and their ranges of validity determined for particular materials.

It is necessary now to seek a more precise expression for the functional $\mathscr{F}$ appearing in (2.1): a number of different forms will be considered. In Section 2 of Chapter 1 it was stated that plastic behaviour associated with a yield stress would not be considered which places a restriction on allowable forms of $\mathscr{F}$. If a material is such that it behaves elastically when the stress is less than a yield value σ_Y and flows when this value is exceeded, then the loading histories 1 and 2 shown in Fig. 1(a) will produce strain responses of the form shown schematically in Fig. 1(b). In this type of situation, which it is desired to exclude from the subsequent discussion, two histories which differ only by a small amount do not necessarily produce responses which also differ by a small amount. The desired restriction can be made by requiring the functional $\mathscr{F}$ to be continuous with respect to its dependence on the strain ϵ. This restriction can now be used to express (2.1) in a more precise form.

B. Integral Forms of Constitutive Relation

Since linear mechanical behaviour is considered in this chapter, $\mathscr{F}$ is a linear continuous functional of its argument function and it can be expressed (Riesz and Nagy, 1955) in the form

$$\sigma(t) = \int_{-\infty}^{t} \zeta(t, \tau)\, \dot{\epsilon}(\tau)\, d\tau \tag{2.2}$$

where $\dot{\epsilon}(\tau) \equiv \partial\epsilon/\partial\tau$ and ζ represents the particular material response function. Equation (2.2) allows for the possibility of material ageing since the stress required to produce a given strain history is not necessarily the same as that required to produce the same strain history at a different time. This is readily seen since if ϵ_1 and ϵ_2 are two strain programmes such that

$$\epsilon_1(t) = f(t), \qquad \epsilon_2(t) = f(t-k) \tag{2.3}$$

then the stresses σ_1 and σ_2 required to produce these strains are given by (2.2) and are such that

$$\sigma_2(t) \neq \sigma_1(t-k) \tag{2.4}$$

in general. If attention is confined to non-ageing or hereditary materials then the special form $\zeta(t, \tau) = \phi(t-\tau)$, so that

$$\sigma(t) = \int_{-\infty}^{t} \phi(t-\tau)\, \dot{\epsilon}(\tau)\, d\tau \tag{2.5}$$

is sufficient to ensure that $\sigma_2(t) = \sigma_1(t-k)$. The function ϕ is known as the stress-relaxation function for the material and it must be determined either by experiment or from the physics of the material structure. When ϕ is a constant and it is assumed that the material was in a virgin unstrained state at $\tau = -\infty$, (2.5) can be integrated to give $\sigma(t) \propto \epsilon(t)$, which is the constitutive relation for an elastic material. The constitutive equation for a Newtonian viscous fluid is obtained by setting

$$\phi(t) = \eta\delta(t) \tag{2.6}$$

where η is the constant viscosity of the fluid and $\delta(t)$ is the Dirac delta function, $\delta(t) = 1$ when $t = 0$; $\delta(t) = 0$ when $t \neq 0$. Thus it is necessary to allow ϕ to contain generalized functions in order to cover the complete spectrum of material behaviour from elastic solids to viscous fluids.

Equation (2.5) is not the only form of linear integral relation that can be written expressing σ in terms of the strain history $\epsilon(\tau)$; alternative relations are

$$\sigma(t) = \int_{-\infty}^{t} \phi_1(t-\tau)\,\epsilon(\tau)\,d\tau$$

$$\sigma(t) = \int_{0}^{\infty} \phi_2(\tau)\frac{d\epsilon}{d\tau}(t-\tau)\,d\tau$$

$$\sigma(t) = \int_{0}^{\infty} \phi_3(\tau)\,\epsilon(t-\tau)\,d\tau \tag{2.7}$$

which can be obtained from (2.5) by transformation of variables and integration by parts. Each of the relations (2.5) and (2.7) is equivalent to an expression of the Boltzmann superposition principle which states that, in this linear system, the total response to a strain history is equivalent to the sum of the responses to a sequence of incremental strain histories.

In the preceding discussion the stress in a material element at a given time has been expressed in terms of the strain history in that element. However, there is no physical reason why one should not consider the dependence of the current strain on the past history of stress and write, by analogy with (2.1),

$$\epsilon(t) = \mathop{\mathscr{F}_1}_{\tau=-\infty}^{t} \{\sigma(\tau)\}. \tag{2.8}$$

The explicit linear form of this relation for a non-ageing viscoelastic material is

$$\epsilon(t) = \int_{-\infty}^{t} \psi(t-\tau)\,\dot{\sigma}(\tau)\,d\tau \tag{2.9}$$

which is the analogue of (2.5), and forms corresponding to (2.7) can also be written. The function ψ is known as the creep function for the material and it must be obtained from experiment or from the physics of the material structure. Since either of (2.5) or (2.9) is sufficient to model the one-dimensional linear behaviour of a material it must, of course, be possible to derive one of the functions ϕ and ψ from a knowledge of the other. This and associated problems will be discussed in Section 3.

C. Differential Form of Constitutive Relation

From a physical interpretation of the stress-relaxation function ϕ appearing in (2.5) (see Chapter 3), and from the results of numerous experimental investigations on real materials, it is apparent that ϕ is a monotonically decreasing function of its argument which does not necessarily, however, tend to zero as the argument tends to infinity. This behaviour also follows from the assumption of fading memory (Christensen, 1971) which states that material response depends more on recent history than earlier events. Thus ϕ may be expressed in the form

$$\phi(t) = \phi_\infty + \int_0^\infty F(\tau)\, e^{-t/\tau}\, d\tau \tag{2.10}$$

where $F(\tau)$ is known as the relaxation spectrum. In principle there is no reason to suppose that $F(\tau)$ should be a function of discrete values:

$$F(\tau) = \sum_{i=1}^{N} A_i\, \delta(\tau - \tau_i) \tag{2.11}$$

where δ is again the Dirac delta. If this is the case then ϕ can be expressed as

$$\phi(t) = \phi_\infty + \sum_{i=1}^{N} A_i\, e^{-t/\tau_i} \tag{2.12}$$

and, although ϕ may not be precisely expressible in this form, it is certainly possible to represent ϕ by an expression of the form (2.12) to an accuracy compatible with that of experimental data. Thus the discrete form (2.11) of relaxation spectrum may be a useful practical approximation. The discrete times τ_i at which (2.11) is non-zero are known as the relaxation times of the material.

For an elastic material ϕ is a constant, and it follows that any relaxation times associated with an elastic material are infinite. For a viscous fluid the relaxation function is $\phi(t) = \eta\delta(t)$ and this can be interpreted by setting $\phi_\infty = 0$ and allowing the relaxation times τ_i to tend to zero.

When the relation (2.12) is substituted into (2.5) there results

$$\sigma(t) = \phi_\infty \epsilon(t) + \sum_{i=1}^{N} A_i e^{-t/\tau_i} \int_{-\infty}^{t} e^{\tau/\tau_i} \dot{\epsilon}(\tau)\, d\tau. \tag{2.13}$$

In the simple special case of a material with only one relaxation time (that is, $N = 1$) this becomes

$$\sigma(t) = \phi_\infty \epsilon(t) + A_1 e^{-t/\tau_1} \int_{-\infty}^{t} e^{\tau/\tau_1} \dot{\epsilon}(\tau)\, d\tau \tag{2.14}$$

and, upon differentiation with respect to t and some manipulation, it can be shown that

$$\tau_1 \dot{\sigma}(t) + \sigma(t) = \phi_\infty \epsilon(t) + \tau_1(\phi_\infty + A_1)\dot{\epsilon}(t). \tag{2.15}$$

In general, when $N > 1$, a similar but more tedious calculation leads to an equation of the form

$$\left[p_0 + p_1 \frac{\partial}{\partial t} + p_2 \frac{\partial^2}{\partial t^2} + \cdots\right]\sigma(t) = \left[q_0 + q_1 \frac{\partial}{\partial t} + q_2 \frac{\partial^2}{\partial t^2} + \cdots\right]\epsilon(t) \tag{2.16}$$

where the p_α and q_α are functions of ϕ_∞, A_i and τ_i. For convenience (2.16) may be written in the form

$$P(D)\,\sigma(t) = Q(D)\,\epsilon(t), \qquad D = \frac{\partial}{\partial t} \tag{2.17}$$

where the operators P and Q are polynomials in D such that D^n is interpreted as $\partial^n/\partial t^n$.

Equations of the form (2.16) are special restricted forms of constitutive relation which depend upon the assumption of a discrete relaxation spectrum or an approximation to the stress-relaxation function ϕ. The relations involve only the current values of stress and strain and their current time rates. A differential rather than an integral formulation may be preferable in the solution of some problems, but the main reason for the popularity of this formulation is probably due to the connection with spring-and-dashpot models (see Section 4) with which early work in this field was largely concerned.

D. Complex Modulus Formulation

The analysis given in this section is purely formal; for a rigorous treatment the reader is referred to Gurtin and Sternberg (1962). It is convenient to introduce the Fourier transforms $\bar{\sigma}$, $\bar{\epsilon}$ and $\bar{\phi}$ defined by

$$\{\bar{\sigma}(\omega), \bar{\epsilon}(\omega)\} = \int_{-\infty}^{\infty} \{\sigma(t), \epsilon(t)\} e^{-i\omega t} dt$$

$$\bar{\phi}(\omega) = \int_{0}^{\infty} \phi(t) e^{-i\omega t} dt. \tag{2.18}$$

Under the formal substitution $s = i\omega$ these are equivalent to Laplace transforms, and the analysis which follows has greater rigour in the latter case. However, the forms (2.18) are more convenient here since they introduce the frequency-like parameter ω.

After some manipulation the transform of (2.5) yields

$$\bar{\sigma}(\omega) = i\omega\bar{\phi}(\omega)\,\bar{\epsilon}(\omega) \tag{2.19}$$

so that

$$\bar{\sigma}(\omega) = E(i\omega)\,\bar{\epsilon}(\omega) \tag{2.20}$$

where

$$E(i\omega) = i\omega\bar{\phi}(\omega). \tag{2.21}$$

When $E(i\omega) \equiv E =$ constant, the inverse transform of (2.20) can be taken, giving $\sigma = E\epsilon$, so that the material is elastic and E represents the elastic modulus. In general E is a complex function of ω and is called the complex modulus of the viscoelastic material. It is convenient for some applications to write the complex modulus in terms of its real and imaginary parts

$$\begin{aligned} E(i\omega) &= E_1(\omega) + iE_2(\omega) \\ &= E_1(\omega)[1 + i \tan \delta(\omega)] \end{aligned} \tag{2.22}$$

where

$$\tan \delta(\omega) = E_2(\omega)/E_1(\omega). \tag{2.23}$$

The function $\delta(\omega)$ is referred to as the loss angle, and $\tan \delta(\omega)$ is the loss tangent; their physical significance will be demonstrated in Chapter 3.

A rather trivial alternative expression to (2.20) is the relation

$$\bar{\epsilon}(\omega) = J(i\omega)\,\bar{\sigma}(\omega) \tag{2.24}$$

where $J(i\omega)$ is the reciprocal of $E(i\omega)$ and is known as the complex compliance of the material. J may also be expressed as a combination of its real and imaginary parts, but it is convenient to write

$$J(i\omega) = J_1(\omega) - iJ_2(\omega). \tag{2.25}$$

In equation (2.5) it was seen that a single function ϕ of a single variable was sufficient to model linear one-dimensional viscoelastic behaviour. In the formulation given in the present sub-section there are apparently two such functions (for example, E_1 and E_2); in fact one of these functions can, in principle be determined when the other is known (see Section 2).

When the relaxation function has a discrete spectrum an alternative derivation of (2.20) can be given. It follows from (2.18) that

$$\int_{-\infty}^{\infty} \frac{d^n f(t)}{dt^n} e^{-i\omega t} dt = (i\omega)^n \bar{f}(\omega) \tag{2.26}$$

by repeated integration by parts, provided that certain derivatives vanish at the limits of integration. In this case, application of (2.18) to (2.17) leads to the equation

$$P(i\omega)\,\bar{\sigma}(\omega) = Q(i\omega)\,\bar{\epsilon}(\omega) \tag{2.27}$$

which is of the form (2.20) with

$$E(i\omega) = Q(i\omega)/P(i\omega). \tag{2.28}$$

2. INTERRELATIONS BETWEEN MATERIAL FUNCTIONS

It has been seen that a single function, for example the stress relaxation function, is sufficient to completely define the one-dimensional linear behaviour of a viscoelastic solid. It follows that all other material functions appearing in alternative formulations of the theory must be capable, in principle, of determination from that function. In some cases determination of a particular function may best be achieved by first calculating an intermediate function; thus interrelations between all of the material functions should be considered.

Since the transform of a constant k is formally equal to $k/i\omega$, (2.21) can be expressed as

$$E(i\omega) - \phi_\infty = i\omega(\bar{\phi} - \bar{\phi}_\infty) = i\omega \int_0^\infty [\phi(t) - \phi_\infty]\, e^{-i\omega t} dt \tag{2.29}$$

and, taking real and imaginary parts, it follows that

$$E_1(\omega) = \phi_\infty + \omega \int_0^\infty [\phi(t) - \phi_\infty] \sin \omega t \, dt$$

$$E_2(\omega) = \omega \int_0^\infty [\phi(t) - \phi_\infty] \cos \omega t \, dt. \tag{2.30}$$

Since $\phi(t)$ is a monotonically decreasing function, tending to the limit ϕ_∞ as $t \to \infty$, the integrals in (2.30) are convergent.

The expressions (2.30) are Fourier sine and cosine transforms and can be inverted to give expressions for $\phi(t)$ in terms of E. When use is made of the result $\phi_\infty = E_1(0) = E_0$, which follows from (2.30) with $\omega = 0$, these relations can be written in the form

$$\phi(t) = E_0 + \frac{2}{\pi} \int_0^\infty \omega^{-1} [E_1(\omega) - E_0] \sin \omega t \, d\omega$$

$$\phi(t) = E_0 + \frac{2}{\pi} \int_0^\infty \omega^{-1} E_2(\omega) \cos \omega t \, d\omega. \tag{2.31}$$

If one of the functions E_1 or E_2 is known, the function ϕ can be determined in principle from (2.31), and then the other component of E can be determined from (2.30). Alternatively E_1 and E_2 are directly related through the equation

$$E_1(\omega) - E_1(0) = \frac{2}{\pi} \int_0^\infty \frac{E_2(\alpha)\,\omega^2}{\alpha(\omega^2 - \alpha^2)} \, d\alpha. \tag{2.32}$$

In a similar manner relations can be obtained connecting the complex compliance J and the creep function ψ. It will be seen in Chapter 3 that ψ is a monotonically increasing function which may not tend to a finite value or slope as $t \to \infty$. For present purposes it is convenient to introduce quantities η and ψ_∞ such that

$$\left.\begin{array}{l} d\psi/dt \to 1/\eta \\ \psi - t/\eta \to \psi_\infty \end{array}\right\} \text{as } t \to \infty. \tag{2.33}$$

The relation $J(i\omega) = i\omega\bar{\psi}$ is now written in the form

$$\begin{aligned} J(i\omega) - \psi_\infty + i/\eta\omega &= i\omega(\bar{\psi} - \bar{\psi}_\infty - \bar{t}/\eta) \\ &= i\omega \int_0^\infty [\psi(t) - \psi_\infty - t/\eta]\, e^{-i\omega t} \, dt \end{aligned} \tag{2.34}$$

where use has been made of the formal result $\bar{t} = -1/\omega^2$. Taking real and imaginary parts of (2.38), and noting the definition (2.28), yields

$$J_1(\omega) = \psi_\infty + \omega \int_0^\infty [\psi(t) - \psi_\infty - t/\eta] \sin \omega t \, dt$$

$$J_2(\omega) = 1/\eta\omega - \omega \int_0^\infty [\psi(t) - \psi_\infty - t/\eta] \cos \omega t \, dt. \tag{2.35}$$

Because of the definitions (2.33) the integrals in (2.35) are convergent and it follows that $J_1(0) = \psi_\infty$ whereas $J_2(0)$ is infinite. It can be shown (see Ferry, 1961) that convergent expressions for the inverses of (2.35) are

$$\psi(t) = J_0 + t/\eta + \frac{2}{\pi} \int_0^\infty \omega^{-1}[J_1(\omega) - J_0] \sin \omega t \, d\omega$$

$$\psi(t) = J_0 + t/\eta + \frac{2}{\pi} \int_0^\infty \omega^{-1}[J_2(\omega) - 1/\eta\omega](1 - \cos \omega t) \, d\omega \tag{2.36}$$

where $J_0 = J_1(0)$.

Equations (2.30) and (2.31) relate stress-relaxation and complex modulus functions; equations (2.35) and (2.36) relate creep and complex compliance functions. These two sets of functions satisfy the equation

$$E(i\omega) = 1/J(i\omega) \tag{2.37}$$

from which it follows that

$$1 + \omega^2 \bar{\phi}\bar{\psi} = 0 \tag{2.38}$$

and

$$E_1(\omega) = \frac{J_1(\omega)}{J_1^2(\omega) + J_2^2(\omega)}, \qquad E_2(\omega) = \frac{J_2(\omega)}{J_1^2(\omega) + J_2^2(\omega)} \tag{2.39}$$

and a similar pair of equations in which the roles of E and J are interchanged. The relations derived above, together with (2.28) and (2.23), are sufficient to determine all of the functions considered here when one of their number is completely known. Further discussion of these interrelations can be found in Ferry (1961), Smith (1958) and Gross (1953).

Many of the results presented above involve the evaluation of integrals between the limits 0 and ∞. If the analytic form of one of the material functions is known, then these integrals can, in principle, be evaluated to give the other material functions. However this is not the case except, for example, for some idealized spring-and-dashpot model (see Section 4). Thus evaluation of the integrals will require numerical treatment using experimental data. In general, for reasons which will be discussed in Chapter 3, this experimental data is known over only a limited range of the integration variable, and evaluations of the integrals can therefore be at best only approximations.

Apart from this practical and computational difficulty, exact evaluation of the interrelations may be tedious, and there are obvious attractions for seeking approximate expressions of a simpler form. In general, however, derivation of such simple forms depends on an assumption, which may be based on experimental data, that a material function has a particularly simple dependence on its argument. For some materials and applications it is adequate to suppose that the loss tangent has a non-zero value which is independent of frequency. Provided that E_1 does not vary rapidly with ω it follows (Kolsky, 1960) that

$$E_2(\omega) \simeq (\pi\omega/2)\, dE_1/d\omega$$

$$E_1(\omega) \simeq E_1(\omega_0) \exp\left[(2/\pi) \tan\delta \log(\omega/\omega_0)\right] \tag{2.40}$$

where ω_0 is a fixed reference frequency. In other suggested approximate forms it is supposed that material functions, or more usually their logarithms, have a linear variation with $\log\omega$ or $\log t$. It is outside the scope of the present work to discuss these approximate forms or their likely areas of validity: the interested reader is referred to Ferry (1961).

3. THREE-DIMENSIONAL THEORY

A linear infinitesimal theory of viscoelasticity for three-dimensional behaviour can be formulated in a manner entirely analogous to the one-dimensional theory discussed in Section 1. With the same qualifying remarks as were made there, the stress components σ_{ij} are supposed to depend on the entire history of the strain components $\epsilon_{ij}(\tau)$ at the same material point. Thus a relationship of the form

$$\sigma_{ij}(t) = \mathop{\mathscr{F}_{ij}}_{\tau=-\infty}^{t} \{\epsilon_{kl}(\tau)\} \tag{2.41}$$

is postulated where $\mathscr{F}_{ij}$ are functionals which, in the general anisotropic case, may be different for all combinations of i and j. In the linear theory (2.41) can be expressed in the form

$$\sigma_{ij}(t) = \int_{-\infty}^{t} c_{ijkl}(t-\tau)\, \dot{\epsilon}_{kl}(\tau)\, d\tau \tag{2.42}$$

where, by analogy with (2.5), c_{ijkl} are stress-relaxation functions for three-dimensional behaviour.

A. Isotropic Materials

For an isotropic material (2.42) takes the simplified form

$$\sigma_{ij}(t) = \delta_{ij} \int_{-\infty}^{t} \lambda(t-\tau)\,\dot{\epsilon}_{kk}(\tau)\,d\tau + 2\int_{-\infty}^{t} \mu(t-\tau)\,\dot{\epsilon}_{ij}(\tau)\,d\tau \tag{2.43}$$

where summation is implied over the repeated suffix. In matrix notation (2.43) can be written as

$$\boldsymbol{\sigma}(t) = \mathbf{I}\int_{-\infty}^{t} \lambda(t-\tau)\,tr\dot{\boldsymbol{\epsilon}}(\tau)\,d\tau + 2\int_{-\infty}^{t} \mu(t-\tau)\dot{\boldsymbol{\epsilon}}(\tau)\,d\tau. \tag{2.44}$$

When the functions λ and μ are constants (2.44) reduces to the form (1.11) for an isotropic elastic solid, λ and μ being the Lamé constants. In (2.44) λ and μ are the two stress-relaxation functions required to describe an isotropic linear viscoelastic solid. If the material can be adequately represented by the assumption of incompressibility, then (2.44) is replaced by

$$\boldsymbol{\sigma}(t) = -p\mathbf{I} + 2\int_{-\infty}^{t} \mu(t-\tau)\dot{\boldsymbol{\epsilon}}(\tau)\,d\tau \tag{2.45}$$

by analogy with (1.13). Again the pressure p is not a material parameter but is an unknown which must be determined from the field equations and the boundary conditions relevant to a particular application. This extra unknown is compensated for by the addition of the continuity relation (1.6), which, in this case, takes the form $\mathrm{tr}\,\boldsymbol{\epsilon} = 0$.

It will be shown in Chapter 4 that equation (2.1) and the physical statement preceding it are somewhat oversimplified, particularly in respect of the choice of a strain measure. Justifications of linear relations like (2.44) are best obtained as special cases of the nonlinear theory (see Section 2B, Chapter 4): for the present the validity of (2.44) can be assumed.

By starting from a relation inverse to (2.41) in which current strain is expressed as a functional of stress history at a point, an alternative constitutive relation is obtained which reduces to the form

$$\boldsymbol{\epsilon}(t) = \mathbf{I}\int_{-\infty}^{t} \theta(t-\tau)\,\mathrm{tr}\,\dot{\boldsymbol{\sigma}}(\tau)\,d\tau + 2\int_{-\infty}^{t} \zeta(t-\tau)\dot{\boldsymbol{\sigma}}(\tau)\,d\tau \tag{2.46}$$

for an isotropic material. In this relation θ and ζ are the two creep functions required to describe isotropic linear behaviour. Alternative relations involving complex modulus and complex compliance functions are readily obtained by applying the transforms (2.18) to (2.44) and (2.46):

$$\bar{\boldsymbol{\sigma}}(i\omega) = \mathbf{I}[i\omega\bar{\lambda}(i\omega)]\,\mathrm{tr}\,\bar{\boldsymbol{\epsilon}}(i\omega) + 2[i\omega\bar{\mu}(i\omega)]\,\bar{\boldsymbol{\epsilon}}(i\omega) \tag{2.47}$$

$$\bar{\boldsymbol{\epsilon}}(i\omega) = \mathbf{I}[i\omega\bar{\theta}(i\omega)]\,\mathrm{tr}\,\bar{\boldsymbol{\sigma}}(i\omega) + 2[i\omega\bar{\zeta}(i\omega)]\,\bar{\boldsymbol{\sigma}}(i\omega). \tag{2.48}$$

By analogy with the transforms of (1.12) and its dual expression for $\boldsymbol{\epsilon}$ in terms of $\boldsymbol{\sigma}$, and by analogy with (2.21), the bracketed expressions in (2.47) are complex moduli functions whereas those in (2.48) are complex compliances.

B. Special Deformations and Interrelations

Interrelations between the pairs of creep, stress-relaxation, modulus and compliance functions are best obtained by considering special one-dimensional cases of the above expressions and then using results derived in Section 2. It is convenient to introduce the deviatoric stress matrix $\mathbf{s}$ and deviatoric strain matrix $\mathbf{e}$ defined by

$$\mathbf{s} = \boldsymbol{\sigma} - \tfrac{1}{3}\mathbf{I}\sigma_0, \qquad \mathbf{e} = \boldsymbol{\epsilon} - \tfrac{1}{3}\mathbf{I}\epsilon_0, \qquad \sigma_0 = \operatorname{tr}\boldsymbol{\sigma}, \qquad \epsilon_0 = \operatorname{tr}\boldsymbol{\epsilon}. \tag{2.49}$$

It is evident that $\operatorname{tr}\mathbf{s} \equiv 0$, $\operatorname{tr}\mathbf{e} \equiv 0$. Using (2.44), (2.46) and (2.49) it follows that

$$\mathbf{s}(t) = 2\int_{-\infty}^{t} \mu(t-\tau)\,\dot{\mathbf{e}}(\tau)\,d\tau \tag{2.50}$$

$$\mathbf{e}(t) = 2\int_{-\infty}^{t} \zeta(t-\tau)\,\dot{\mathbf{s}}(\tau)\,d\tau \tag{2.51}$$

$$\sigma_0(t) = \int_{-\infty}^{t} [3\lambda(t-\tau) + 2\mu(t-\tau)]\,\dot{\epsilon}_0(\tau)\,d\tau \tag{2.52}$$

$$\epsilon_0(t) = \int_{-\infty}^{t} [3\theta(t-\tau) + 2\zeta(t-\tau)]\,\dot{\sigma}_0(\tau)\,d\tau. \tag{2.53}$$

In a situation involving only shear in the X_1, X_2 plane, the only non-zero components of $\boldsymbol{\sigma}$ and $\boldsymbol{\epsilon}$ are σ_{12} and ϵ_{12}, and the only non-zero components of $\mathbf{s}$ and $\mathbf{e}$ are s_{12} and e_{12}. Equations (2.50) and (2.51) are of the form (2.5) and (2.9) in which σ, ϵ, ϕ, ψ are replaced by s_{12}, e_{12}, 2μ, 2ζ respectively. It therefore follows from (2.38) that

$$1 + 4\omega^2\,\bar{\mu}\bar{\zeta} = 0 \tag{2.54}$$

which relates the shear creep and stress-relaxation functions.

When the loading of the material takes the form of a hydrostatic pressure $p(t)$ the non-zero components of $\boldsymbol{\sigma}$ and $\boldsymbol{\epsilon}$ are

$$\sigma_{11} = \sigma_{22} = \sigma_{33} = -p, \qquad \epsilon_{11} = \epsilon_{22} = \epsilon_{33} = \epsilon \tag{2.55}$$

and it follows that

$$\sigma_0(t) = -3p(t), \qquad \epsilon_0(t) = 3\epsilon(t). \tag{2.56}$$

Equations (2.52) and (2.53) are now of the form (2.5) and (2.9) with $-3p$, 3ϵ, $3\lambda+2\mu$, $3\theta+2\zeta$ replacing σ, ϵ, ϕ, ψ. It follows from (2.38) that

$$1+\omega^2(3\bar{\lambda}+2\bar{\mu})(3\bar{\theta}+2\bar{\zeta})=0. \tag{2.57}$$

For an elastic material $\lambda+\frac{2}{3}\mu$ is the bulk modulus of the material [see equations (1.13)]; the combination $\lambda(t-\tau)+\frac{2}{3}\mu(t-\tau)$ is the bulk stress-relaxation function for the viscoelastic material. If this function is known (2.57) permits determination of $\theta(t-\tau)+\frac{2}{3}\zeta(t-\tau)$ which is the bulk creep function for the material. Alternatively, if λ and μ are known, ζ can be calculated from (2.54) followed by θ from (2.57).

Finally the situation is considered in which σ_{11} is the only non-zero stress component as, for example, in the uni-directional loading of a rod. In this case the non-zero strain components are ϵ_{11} and $\epsilon_{22}=\epsilon_{33}$, and the transform of equation (2.44) yields

$$\begin{aligned}\bar{\sigma}_{11}&=i\omega\{\bar{\lambda}(\bar{\epsilon}_{11}+2\bar{\epsilon}_{22})+2\bar{\mu}\bar{\epsilon}_{11}\}\\ 0&=i\omega\{\bar{\lambda}(\bar{\epsilon}_{11}+2\bar{\epsilon}_{22})+2\bar{\mu}\bar{\epsilon}_{22}\}\end{aligned} \tag{2.58}$$

and it follows that

$$\bar{\sigma}_{11}=i\omega[(3\bar{\lambda}+2\bar{\mu})/(\bar{\lambda}+\bar{\mu})]\,\bar{\mu}\bar{\epsilon}_{11}=E_*(i\omega)\,\bar{\epsilon}_{11} \tag{2.59}$$

$$\bar{\epsilon}_{22}=-[\bar{\lambda}/2(\bar{\lambda}+\bar{\mu})]\,\bar{\epsilon}_{11}=-\nu(i\omega)\,\bar{\epsilon}_{11}. \tag{2.60}$$

When λ and μ are constants $\bar{\lambda}=\lambda/i\omega$, $\bar{\mu}=\mu/i\omega$ and E_* and ν reduce to the Young's modulus and Poisson's ratio for an elastic material [see equations (1.13)]. The functions $E_*(i\omega)$ and $\nu(i\omega)$ defined in (2.59) and (2.60) are the complex Young's modulus and complex Poisson's ratio for a viscoelastic material. In a similar manner it is seen that $i\omega\bar{\mu}$ and $i\omega(\bar{\lambda}+\frac{2}{3}\bar{\mu})$ are complex shear and bulk moduli respectively. Complex compliance functions are also readily derived.

The transform of the constitutive relation (2.44) may be written in the form

$$\bar{\boldsymbol{\sigma}}=\mathbf{I}L\,\mathrm{tr}\,\bar{\boldsymbol{\epsilon}}+2G\bar{\boldsymbol{\epsilon}} \tag{2.61}$$

where $L=i\omega\bar{\lambda}$ and $G=i\omega\bar{\mu}$ are complex modulus functions corresponding to λ and μ. Equation (2.61) is identical with the transform of the elastic relation (1.12) if the elastic constants λ and μ are replaced by L and G respectively. In addition the transformed field equations are identical for elastic and viscoelastic materials. Provided that the boundary conditions of a particular problem do not change their *form* with passage of time, they too can be transformed and do not depend on whether the material is viscoelastic or elastic. In these circumstances the entire set of transformed equations has the same form for the two types of material behaviour, the quantity ω appearing simply as a constant parameter. This fact permits a statement of the viscoelastic correspondence principle or elastic–viscoelastic analogy. Under the circumstances discussed above the solution of a boundary value problem for a viscoelastic material can be obtained by transforming the solution of the corresponding elastic problem, replacing elastic constants by the relevant complex moduli, and then performing the inverse transformations. In principle a great number of solutions to viscoelastic problems can be generated in this way from known elastic solutions. The most difficult part of this operation is the final inverse transform since the material parameters are functions of the integration variable.

C. Anisotropic Materials

In Section 5 of Chapter 1 it was noted that the elastic moduli c_{ijkl} satisfy the symmetry relations (1.10) so that only 21 parameters are involved in the most general case. For the anisotropic viscoelastic relation (2.42) symmetry of the stress and strain matrices requires that

$$c_{ijkl}(\tau) = c_{jikl}(\tau) = c_{ijlk}(\tau) \tag{2.62}$$

which restricts the number of independent functions to 36. For an elastic material the reduction to 21 independent parameters is effected by recourse to a thermodynamic argument which leads to the existence of a strain energy function and which results in the additional restriction $c_{ijkl} = c_{klij}$. For a viscoelastic material this argument does not remain valid and the restrictions (2.62) alone can be applied for all times. Gurtin and Herrera (1965) have shown that

$$c_{ijkl}(0) = c_{klij}(0), \qquad c_{ijkl}(\infty) = c_{klij}(\infty) \tag{2.63}$$

under the assumption of linear isothermal viscoelasticity. Thus the short and long time material behaviour satisfies the elastic restrictions.

In principle anisotropic formulations involving creep, complex modulus or complex compliance functions can be derived, but they will not be discussed here.

4. SPRING-AND-DASHPOT MODELS

In the early development of viscoelasticity theory much use was made of the modelling of materials by networks made up of simple elastic and viscous elements. The elastic element is a spring, shown schematically in Fig. 2(a), for which stress and strain are connected by the relation $\sigma = E\epsilon$, where E is the elastic modulus. The simple viscous element is represented by a dashpot [shown in Fig. 2(b)] which is merely a piston moving in a viscous fluid contained in a cylindrical container. For this element $\sigma = \eta\dot{\epsilon}$, where η is the viscosity of the element.

Since viscoelastic materials combine elastic and viscous effects, it is possible to model them approximately by combining the simple components in various parallel and series arrangements; three simple combinations are shown in Fig. 3. The constitutive relations appropriate to these models will now be derived. The model shown in Fig. 3(a) is known as the Maxwell model and, since the

Fig. 2. Network elements: (a) spring, (b) dashpot.

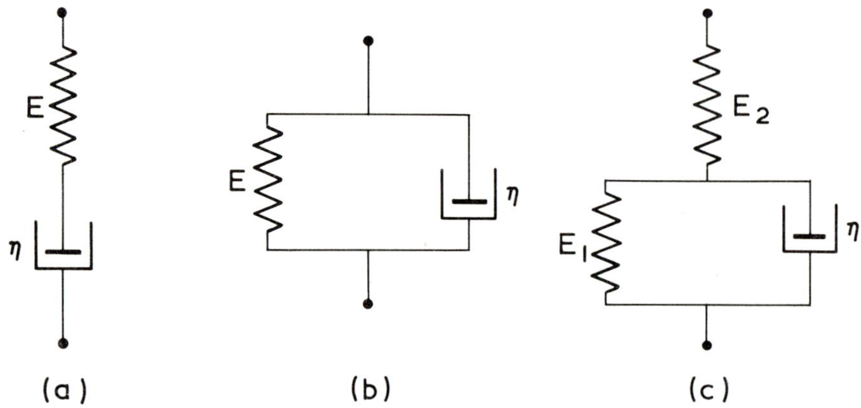

Fig. 3. Simple spring-and-dashpot models: (a) Maxwell, (b) Voigt, (c) standard linear solid.

elements are connected in series, the stress in the model is equal to the stress in each element, whereas the strain is the sum of the strains in the two elements, that is

$$\sigma = \sigma_E = \sigma_\eta, \qquad \epsilon = \epsilon_E + \epsilon_\eta \tag{2.64}$$

where suffixes E and η denote components associated with the elastic and viscous elements respectively. Combining (2.64) with the constitutive relations

$$\sigma_E = E\epsilon_E, \qquad \sigma_\eta = \eta\dot{\epsilon}_\eta \tag{2.65}$$

it readily follows that

$$\frac{\sigma}{\eta} + \frac{\dot{\sigma}}{E} = \dot{\epsilon} \tag{2.66}$$

which is the constitutive relation for the Maxwell model. In a similar manner it is readily shown that the constitutive relation for the Voigt model shown in Fig. 3(b) is

$$\sigma = E\epsilon + \eta\dot{\epsilon} \tag{2.67}$$

and that for the *standard linear solid* shown in Fig. 3(c) is

$$(1 + E_1/E_2)\,\sigma + (\eta/E_2)\,\dot{\sigma} = E_1\,\epsilon + \eta\dot{\epsilon}. \tag{2.68}$$

All the equations (2.66)–(2.68) are special cases of the differential formulation (2.16) and spring-and-dashpot models can be devised to model any relation of this form. However, the main value of spring-and-dashpot models is in providing a relatively simple qualitative picture of the manner in which viscoelastic materials behave, and this is only feasible when the models contain only a few elements. From a quantitative point of view these models have little value (see Chapter 3, Section 4) although some authors have attempted to elevate them to a higher status.

It may be noted in passing that the relation (2.68) for a standard linear solid is identical with (2.15) for a material with only one relaxation time provided that

$$\tau_1 = \eta/(E_1 + E_2), \qquad \phi_\infty = E_1\,E_2/(E_1 + E_2), \qquad A_1 = E_2^2/(E_1 + E_2). \tag{2.69}$$

This correspondence exists since the standard linear solid is the most general model containing only one dashpot.

For these simple models the material parameters associated with alternative constitutive relations are readily determined by using the interrelations derived in Section 2. In the case of a Maxwell model comparison of (2.16) and (2.66) yields

$$\begin{aligned} &p_0 = 1/\eta, \qquad p_1 = 1/E, \qquad p_i = 0, \qquad i > 1; \\ &q_1 = 1, \qquad q_i = 0, \qquad i \neq 1. \end{aligned} \tag{2.70}$$

Thus it follows from (2.28) that for a Maxwell model

$$E(i\omega) = i\omega/[1/\eta + i\omega/E] \tag{2.71}$$

and

$$\begin{aligned} E_1(\omega) &= \omega^2\,E^{-1}\left[\frac{1}{\eta^2} + \frac{\omega^2}{E^2}\right]^{-1} \\ E_2(\omega) &= \omega\eta^{-1}\left[\frac{1}{\eta^2} + \frac{\omega^2}{E^2}\right]^{-1} \\ \tan\delta(\omega) &= E/\omega\eta. \end{aligned} \tag{2.72}$$

Hence by (2.31) and (2.38)

$$\phi(t) = E\exp(-Et/\eta), \qquad \psi(t) = 1/E + t/\eta. \tag{2.73}$$

Finally the dual of relations (2.39) show that

$$J_1(\omega) = 1/E, \qquad J_2(\omega) = 1/\eta\omega, \tag{2.74}$$

which completes the determination of material parameters for the Maxwell model.

In a similar manner it can be shown that the material parameters appropriate to the Voigt model are

$$p_0 = 1, \qquad q_0 = E, \qquad q_1 = \eta, \qquad (p_i = q_i = 0 \text{ otherwise})$$

$$E_1(\omega) = E, \qquad E_2(\omega) = \eta\omega, \qquad \tan\delta(\omega) = \eta\omega/E,$$

$$J_1(\omega) = E(E^2 + \eta^2\omega^2)^{-1}, \qquad J_2(\omega) = \eta\omega(E^2 + \eta^2\omega^2)^{-1},$$

$$\phi(t) = E + \eta\delta(t), \qquad \psi(t) = \frac{1}{E}[1 - \exp(-Et/\eta)]. \tag{2.75}$$

The corresponding results for a standard linear solid are readily determined. In particular

$$\phi(t) = \frac{E_2}{E_1 + E_2}\left\{E_1 + E_2\exp\left(-\frac{E_1 + E_2}{\eta}t\right)\right\},$$

$$\psi(t) = \frac{1}{E_1 E_2}\{E_1 + E_2[1 - \exp(-E_1 t/\eta)]\}. \tag{2.76}$$

Explicit forms of constitutive relation for more general spring-and-dashpot models can be determined but evaluation of interrelations between the various material functions becomes rather tedious.

Chapter 3 *Experimental Determination of Linear Properties*

In principle the direct experimental determination of material functions for a linear isotropic viscoelastic material is straightforward. However certain difficulties arise in practice which make it necessary in some circumstances to modify these direct procedures. Section 1 considers the possible direct methods; alternative techniques are given in Section 2 and some of the experimental considerations which govern choice of method are discussed in Section 3. The following section presents some information about the typical behaviour of polymers in both the glassy and rubbery states and a discussion of the relevance of simple spring-and-dashpot models to the representation of real materials. Variations of material properties with changes in temperature are discussed briefly in Section 5 and a brief account of material property determination for anisotropic materials is given in the final section.

1. DIRECT EXPERIMENTAL DETERMINATION OF MATERIAL FUNCTIONS

It will be assumed in the present section that the material is subjected at all times to a homogeneous stress and strain field: thus $\boldsymbol{\sigma}$ and $\boldsymbol{\epsilon}$ depend only on time and this dependence will take different forms in each of the proposed experiments. It follows from the discussion in Chapter 2, Section 3 that material functions for three-dimensional behaviour may be determined from experiments which are governed by equations which are essentially one-dimensional in character. If, for example, separate experiments are performed to determine the stress-relaxation functions for shear and bulk behaviour, then Section 3B of Chapter 2 shows that the functions μ and $\lambda + \frac{2}{3}\mu$ can be determined. Thus λ and μ are both known, and general three-dimensional behaviour is characterized (provided, of course, that the material does exhibit linear isotropic viscoelastic behaviour). Similar remarks apply if experiments are performed to determine creep, complex modulus or complex compliance functions. As an alternative to either the shear or bulk experiment, or as an additional check, experiments involving other special deformations can be considered (for example, uniaxial extension or beam bending).

Because of the facts discussed above, it is sufficient to consider the problem of experimental determination of material functions occurring in the one-dimensional theory in Chapter 2, Section 1. For purposes of the present section it is not necessary to specify the nature of the one-dimensional deformation, since the main interest here is in general principles; however, this is sometimes necessary in the discussion of experimental considerations given in Section 3.

A. Creep Test

In this test the material specimen, which is initially supposed to be in a virgin state, is subjected to a load which is applied as quickly as possible and varies in such a manner as to keep the stress at a constant value after load application. The resulting deformation or strain in the specimen is measured as a function of time elapsed since the load was first applied. If the time of load-application is chosen as the origin of the time scale, the stress in the specimen has the form

$$\sigma(t) = \sigma_0 H(t) \tag{3.1}$$

where σ_0 is a constant depending on the magnitude of the applied load and the dimensions of the specimen; $H(t)$ denotes the Heaviside step function $H(t) = 0$, $t < 0$; $H(t) = 1$, $t \geqslant 0$. Since $\dot{H}(t) = \delta(t)$, where δ is the Dirac delta function, and

$$\int_a^b f(x)\,\delta(x-c)\,dx = f(c), \qquad a \leqslant c \leqslant b \tag{3.2}$$

equation (2.9) can be used to write the strain response to the stress (3.1) in the form

$$\epsilon(t) = \sigma_0\,\psi(t). \tag{3.3}$$

Thus the creep function $\psi(t)$ is readily determined by dividing the measured strain response by the known constant σ_0.

The general features of the creep function for real materials are shown schematically in Fig. 1. Provided that inertia effects are negligible the strain in an elastic material immediately attains some value and then remains constant. For a viscous fluid the material flows at a constant rate, so that ψ is proportional to t. These remarks are consistent with the requirements that $\psi = 1/E$, $\psi = t/\eta$ in order to obtain the constitutive relations for elastic and viscous materials. For a viscoelastic material there is a relatively rapid increase in ψ for small values of t, that is, soon after application of the load, which may or may not contain an instantaneous elastic component. As t increases the slope of the curve decreases, and as $t \to \infty$ the slope may approach zero or a finite non-zero value. The former, which implies a final equilibrium state, is appropriate to the behaviour of crosslinked polymers; the latter behaviour would be expected from an un-crosslinked polymer. These possible forms of response were allowed for in equations (2.33) *et seq*. Further discussion of the form of the creep functions for real materials will be given in Section 4.

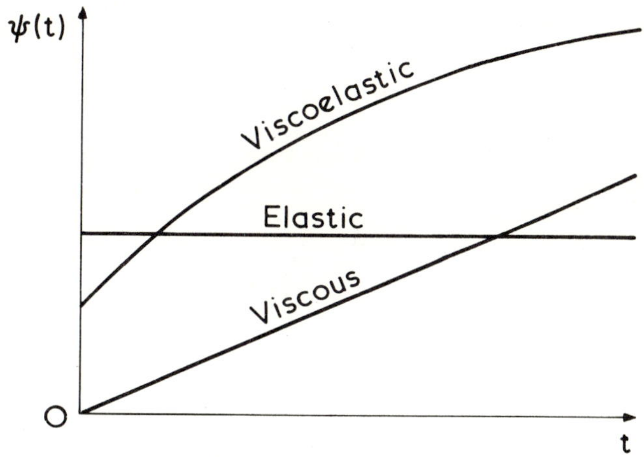

Fig. 1. Schematic form of creep function $\psi(t)$.

B. Stress-Relaxation Test

This is the dual of the creep test: the material specimen is now subjected at time $t = 0$ to a suddenly applied strain which is maintained constant thereafter. Thus the applied strain has the form

$$\epsilon(t) = \epsilon_0 H(t) \tag{3.4}$$

where ϵ_0 is a constant, and the stress required to maintain this strain is given by (2.5) to be

$$\sigma(t) = \epsilon_0 \phi(t). \tag{3.5}$$

If $\sigma(t)$ is measured throughout the experiment, the stress-relaxation function $\phi(t)$ is readily determined from (3.5). It can be seen that $\phi(t)$ is merely the stress required to maintain unit strain and, due to relaxation processes in viscoelastic material, this is a decreasing function of time (see Fig. 2). For an elastic material ϕ is a constant. A crosslinked polymer exhibits long-term elastic behaviour exemplified by an equilibrium stressed state and, in this case ϕ tends to a finite non-zero value as $t \to \infty$. In an un-crosslinked polymer the relaxation processes may continue until the stress completely vanishes: that is, until a stress is no longer required to maintain the strain in the material. Here $\phi \to 0$ as $t \to \infty$. A special case of this latter behaviour is that of a viscous fluid for which $\phi(t) = \eta\delta(t)$; thus a stress is required instantaneously but immediately relaxes to zero. In a viscoelastic material the elastic mechanisms delay this relaxation.

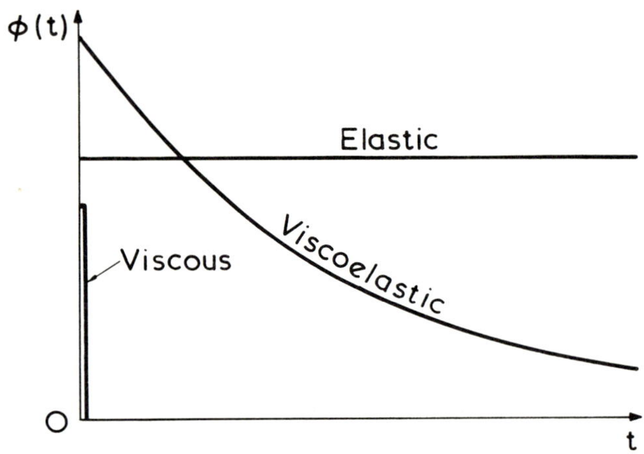

Fig. 2. Schematic form of stress-relaxation function $\phi(t)$.

A common variant on the stress-relaxation test is the constant-rate test in which the material specimen is in a virgin state up to $t = 0$, and is then subjected to a constant strain rate, thus

$$\dot{\epsilon}(t) = \bar{\epsilon} H(t). \tag{3.6}$$

In this case (2.5) gives

$$\sigma(t) = \bar{\epsilon} \int_0^t \phi(t - \tau)\, d\tau \tag{3.7}$$

so that

$$\phi(t) = \frac{1}{\bar{\epsilon}} \frac{d\sigma}{dt} \tag{3.8}$$

Although a constant-rate test has some experimental advantages over a stress-relaxation test, it should be noted that (3.8) requires derivatives of experimental data with consequent problems of maintaining accuracy.

C. Forced Vibration

In this test the material specimen is forced to execute homogeneous vibrations at some frequency Ω. It is immaterial whether stress or strain is the applied quantity provided that the other is measured. For viscoelastic materials, not only will the amplitudes of stress and strain be different, but they will also be out of phase. Thus stress and strain can be written in the complex forms

$$\sigma(t) = \sigma_0 \exp(i\Omega t), \qquad \epsilon(t) = \epsilon_0 \exp[i(\Omega t - \delta)] \tag{3.9}$$

where physical quantities are the real parts of these expressions and δ denotes a phase lag. Substitution of (3.9) into (2.17) yields

$$P(i\Omega)\,\sigma_0 \exp(i\Omega t) = Q(i\Omega)\,\epsilon_0 \exp[i(\Omega t - \delta)] \tag{3.10}$$

so that, by making use of (2.28),

$$E(i\Omega) = (\sigma_0/\epsilon_0)\exp(i\delta). \tag{3.11}$$

The same result is obtained by formal substitution of (3.9) into (2.20). From (3.11) it follows that

$$E_1(\Omega) = \frac{\sigma_0}{\epsilon_0}\cos\delta, \qquad E_2(\Omega) = \frac{\sigma_0}{\epsilon_0}\sin\delta \tag{3.12}$$

and $E_2/E_1 = \tan\delta$. This justifies use of the same symbol in (3.9) and (2.23) and provides a physical interpretation of the loss angle. In a similar manner it can be shown that

$$J_1(\Omega) = \frac{\epsilon_0}{\sigma_0}\cos\delta, \qquad J_2(\Omega) = \frac{\epsilon_0}{\sigma_0}\sin\delta. \tag{3.13}$$

Thus it is possible in a single experiment at frequency Ω to measure the loss angle $\delta(\Omega)$ directly, and to deduce the values of the modulus and compliance at $\omega = \Omega$ from (3.12) and (3.13). By repeating the experiment at other frequencies, these functions can be determined over a wide range of their argument. Some typical curves of modulus and compliance function will be demonstrated and discussed in Section 4.

2. ALTERNATIVE METHODS FOR DETERMINING COMPLEX MODULUS

In each of the three fundamental methods described above stress and strain have been considered to be the input and output functions, and it has been assumed that inertia effects are negligible. In practice creep and stress-relaxation results are influenced by inertia over a short period of time following application of the input function. These practical difficulties are discussed further in Section 3. For the complex moduli measurements discussed above the analysis is sufficient provided that the applied frequency Ω is not close to one of the natural frequencies of the mechanical system and that the specimen dimensions are small compared with the wavelength of mechanical waves with characteristic frequency Ω. Three further experimental techniques are now considered where inertia effects are an essential ingredient of the method.

A. Resonance Measurements

The basic principles of this method are most simply demonstrated through the simple extensional experiment illustrated in Fig. 3. The specimen is assumed to be in the form of a rod of initial length l and cross-sectional area A. It is built into a rigid support at one end, and has a mass M attached at the other

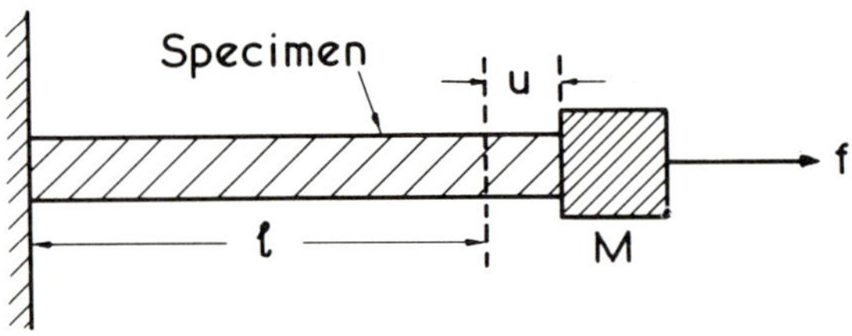

Fig. 3. Schematic diagram of simple resonance experiment in extension.

end. The system is set in motion by an applied force f as shown; f varies sinusoidally with time and has frequency Ω. The force f and the extension of the specimen u are out of phase:

$$u = u_0 \exp(i\Omega t), \qquad f = f_0 \exp(i\Omega t + i\theta) \tag{3.14}$$

where θ is the phase angle.

Provided that the inertia of the specimen can be neglected its effect on the mass M can be considered to be that of a simple viscoelastic spring. The equation of motion of the mass M is readily seen to be

$$f_0 \exp i\theta - AEu_0/l = -Mu_0\,\Omega^2 \tag{3.15}$$

where E is the complex modulus evaluated at frequency Ω, and use has been made of (2.5) and (2.21). Separating (3.15) into real and imaginary parts and eliminating θ between the resulting equations leads to the expression

$$u_0 = f_0\, l[(AE_1 - Ml\Omega^2)^2 + A^2 E_2^2]^{-1/2}. \tag{3.16}$$

When the local variation of E_1 and E_2 with Ω is neglected, inspection of (3.16) shows that u_0 has a maximum value of $u_0 = f_0 l/AE_2$ which is attained when Ω takes its resonant value Ω_R, where

$$\Omega_R = (E_1 A/Ml)^{1/2}. \tag{3.17}$$

In the resonance experiment the frequency Ω of the exciting force f is varied until the amplitude of the vibration is a maximum: from a knowledge of this resonant frequency the value of E_1 can be determined from (3.17), since the other quantities are known. In order to determine the value of E_2 the frequencies Ω_1 and Ω_2 ($\Omega_1 > \Omega_2$) are measured at which, for example, the amplitude of vibration is just one-half of that at Ω_R. These values are obtained by setting $u_0 = f_0 l/2AE_2$ in (3.16), and it is readily shown that

$$[\Omega_1 - \Omega_2]/\Omega_R = [1 + \sqrt{3}E_2/E_1]^{1/2} - [1 - \sqrt{3}E_2/E_1]^{1/2} \tag{3.18}$$

from which E_2 can be determined. Since $E_2/E_1 \ll 1$ for most real materials, it follows that

$$E_2 \simeq E_1[\Omega_1 - \Omega_2]/\sqrt{3}\Omega_R \tag{3.19}$$

from which E_2 can be determined.

Derivation of equations (3.17) and (3.19) for E_1 and E_2 is dependent upon the assumption that these material functions are approximately independent of Ω within the frequency range $\Omega_2 \leqslant \Omega \leqslant \Omega_1$. This is a reasonable approximation for most real materials and the assumption is usually justifiable. In order to evaluate $E_1(\omega)$ and $E_2(\omega)$ at other values of ω the value of M may be changed to obtain different resonant frequencies. Alternatively some variation in resonant frequency may be achieved by changing the dimensions of the specimen. When the mass M has a small value it is necessary to include the inertia of the specimen in the equation of motion of the system; this is particularly relevant to the cases where no added masses are employed. For further details of these methods the reader is referred to Ferry (1961).

B. Free Vibration

The basic principles of this method may be demonstrated by reference to the simple experimental situation of Fig 3. In this case $f \equiv 0$ and the system is set in motion by an initial displacement. Since a viscoelastic material produces damping of the ensuing vibrations, the displacement u can be written as

$$u = u_0 \exp(-\alpha t + i\Omega t) \tag{3.20}$$

where Ω is the natural frequency of vibration of the system and α is a constant. In this case the equation of motion of the mass M gives

$$-AE/l = (i\Omega - \alpha)^2 M \tag{3.21}$$

where E is evaluated at frequency Ω, and it follows that

$$AE_1/lM = \Omega^2 - \alpha^2, \qquad AE_2/lM = 2\Omega\alpha. \tag{3.22}$$

It is now convenient to define the logarithmic decrement Δ as the natural logarithm of the ratio of the amplitudes of two successive oscillations on the same side of the equilibrium position, so that

$$\Delta = 2\alpha\pi/\Omega \tag{3.23}$$

and it follows that

$$E_1 = (M\Omega^2 l/A)(1 - \Delta^2/4\pi^2)$$
$$E_2 = (M\Omega^2 l/A)\Delta/\pi. \tag{3.24}$$

In the free-vibration experiment Ω and Δ are to be measured and $E_1(\Omega)$ and $E_2(\Omega)$ can then be calculated from (3.24). In many practical situations (3.24) can be replaced by

$$E_1(\Omega) \simeq M\Omega^2 l/A, \qquad \pi E_2(\Omega) \simeq \Delta(\Omega) E_1(\Omega). \tag{3.25}$$

Values of E_1 and E_2 for other argument values can be determined by varying the mass M or specimen dimensions so as to achieve other natural frequencies of vibration. When the experiments are performed in, for example, shear or torsion rather than extension, different moduli are involved. However, the analysis is similar to that given above with other parameters replacing the form factor Ml/A.

It may be noted that the above analysis is only approximate since inertia of the specimen is neglected. In addition, since the Fourier spectrum of the damped harmonic oscillation (3.20) contains frequency components in a narrow band of frequencies around the frequency $\omega = \Omega$, it is strictly only an approximation to evaluate E in (3.21) at the single frequency Ω. In practice these effects are insignificant and the analysis given above is adequate.

C. Wave Propagation

The velocity of propagation of waves travelling in viscoelastic materials depends upon the frequency (or frequency spectrum) of the excitation; since the wavelength depends on velocity and frequency, it follows that wavelength is also frequency dependent. When the frequency is sufficiently high the wavelength may become comparable with the dimensions of the specimens used in the free vibration and resonance experiments. At these high frequencies standing-wave analyses are not appropriate, and it is necessary to consider instead methods employing travelling waves.

For the sake of definiteness the case is considered of extensional waves propagating in a thin fibre of material; the analysis is appropriate to other situations with a simple change of notation. The displacement at a typical point in the fibre is written as

$$u(x, t) = u_0 \exp(-\alpha x) \exp[-i(kx - \Omega t)] \tag{3.26}$$

where x denotes distance along the fibre from the point ($x = 0$) at which the excitation

$$u(0, t) = u_0 \exp(i\Omega t) \tag{3.27}$$

is applied. If the displacement at some station ($x = x_1$) in the fibre is recorded, the ratio of the amplitudes of vibration at $x = x_1$ and $x = 0$ yields the value of the attenuation factor α. Similarly, comparison of the phases at $x = x_1$ and $x = 0$ yields the value of k. These measured values of k and α, which apply to the particular frequency Ω, can be used to determine $E_1(\Omega)$ and $E_2(\Omega)$. Using (3.26), the equation of motion at a typical point in the fibre gives

$$E(ik + \alpha)^2 = -\rho\Omega^2 \tag{3.28}$$

where ρ denotes material density and E is evaluated at frequency Ω. It follows that

$$(\alpha^2 - k^2) E_1 - 2\alpha k E_2 = -\rho\Omega^2$$

$$2\alpha k E_1 + (\alpha^2 - k^2) E_2 = 0 \tag{3.29}$$

and E_1 and E_2 can be determined from the measured values of α and k. These quantities can be determined for other values of their arguments by performing experiments with different values of Ω.

Relations (3.28) and (3.29) hold for any type of plane viscoelastic wave provided that E is taken to be the appropriate modulus. In the case of extensional vibrations of a fibre E denotes the Young's modulus; for torsional waves in a circular cylinder E is the shear modulus μ, whereas for longitudinal waves of plane strain travelling in a large block of solid (no lateral motion) it corresponds to $k + \frac{4}{3}\mu$.

An alternative to considering the propagation of continuous harmonic vibrations is to consider the propagation of stress-pulses of arbitrary shape. A pulse of short duration can be generated at the end of a rod of material, either by detonation of a small explosive charge or by collision of a projectile, and the passage of the resulting stress wave can be recorded at convenient stations along the rod. By performing a Fourier synthesis of the pulse at two stations, data can be obtained which are similar to those discussed for a continuous harmonic wave train, but which determine material functions over a short but finite range of frequency in a single experiment (see Kolsky, 1966).

3. EXPERIMENTAL CONSIDERATIONS

The purpose of this book is to discuss general principles rather than matters which are of purely experimental or theoretical detail. However, when a theory suggests a particular form of experiment for determining material properties, attention must obviously be given to the conditions under which the experiment can be performed. Sections 1 and 2 have presented a number of possible techniques through which the entire material behaviour may be determined in principle. In practice the suitability of a particular method is governed by the material under consideration, by the particular mode of deformation of interest, and by the range of argument values over which material functions are required. These factors are discussed in general terms in the present section: for greater detail the reader is referred to Ferry (1961).

A. Creep Test

The test described in Section 1A requires application to the specimen of a homogeneous stress field having the time-dependent form (3.1). In practice such a stress field must be applied through appropriate boundary forces; the geometrical configuration and the form of these boundary tractions will depend on the mode of deformation considered. For uniaxial extension the experiment performed is no more than a sophisticated version of the simple test discussed in Chapter 1, Section 2A. The specimen is chosen in the form of a rod or, more usually, a strip, and the load L is related to the stress σ by $L = \sigma A$, where A is the specimen cross-sectional area. During the course of the test A is a time-dependent function; this presents no difficulties if nominal stress (load divided by initial area) is used in the analysis, since both L and σ then have the same time-dependent form, and the conditions of the test can be achieved by applying a load which is constant for $t > 0$. When true stress (load divided by current area) is used

$$L(t) = \sigma A(t) = \sigma_0 H(t) A(t) \tag{3.30}$$

and the applied load must be continuously adjusted as the area of cross-section varies. If the material is approximately incompressible, $A(t)$ is simply related to the extension of the specimen and it is not difficult to arrange a mechanical system which automatically reduces L as A is reduced by extension (see Andrade, 1914).

To ensure that material behaviour remains in the linear range during a test, an upper limit must be placed upon the magnitude of the constant σ_0 to be used. For relatively stiff materials this may result in the need to measure very small strains during the test, with consequent difficulties of achieving good accuracy. For less stiff materials extensions may be much larger and therefore more readily measured, but in this case the distinction between nominal and true stress is more important. Whatever the stiffness of the material, the creep rig must be constructed so that it is much stiffer than the test specimen; this is a condition which is much more easily satisfied for plastics than for metals.

Creep behaviour in shear is most readily measured by considering torsional creep in rods subjected to a suitably applied torque (see Benbow, 1956). Since the dimensions of the specimen do not change during this test, the distinction between nominal and true stress does not arise, and the torque input and twist data are readily related to the stress and strain components considered in Section 1A. Material data characterizing bulk behaviour are hard to obtain due to practical difficulties encountered in attempting to apply hydrostatic pressures and measure volume changes with sufficient accuracy. One method of load application is by immersion of the specimen in a fluid, but compressibility of the fluid and of any trapped air bubbles leads to difficulties of interpretation. For these reasons bulk properties are deduced from data for shear and simple extension. Other loading systems can be employed, for example beam bending (see Ferry, 1961) or biaxial tension of a sheet of material, but the latter method introduces greater experimental difficulties than the two simpler situations already discussed.

In engineering applications the time scales of possible interest can vary from a few microseconds to several years. The particular application which is envisaged will determine the time scale required in the corresponding creep test. However, there will often arise a conflict between what is required by the engineering application and what is attainable from a test. In describing the creep test it has been stated that load application must be made as quickly as possible but in such a way as to make inertia effects insignificant. Automatic damped devices have been developed for this purpose (see Dunn *et al.*, 1964), but it should be recognized that there is always some lower time value below which the condition (3.1) is not achieved. Depending on the sophistication of the apparatus, and the properties of the material, this lower time value may be of the order of several seconds although it can be as low as 0.004 sec (Benbow, 1956). It follows that material behaviour in the low microsecond range cannot be obtained from creep data; fortunately it can be estimated from high-frequency dynamic data and the approximate interrelations mentioned in Chapter 2, Section 2. At the other end of the time scale it is necessary in some engineering design calculations to predict component behaviour after several years of load application. This time scale is beyond that normally studied in routine creep tests, thus the time interval over which creep data is available also has an upper limit. To some extent it is possible that low-frequency dynamic measurements can supplement this lack of data, but for very long times this depends on dynamic data at extremely low frequencies, and similar difficulties apply at the limits of the frequency range.

Finally it is noted that the mechanical properties of polymers are very sensitive to changes in temperature and this requires strict temperature control if data of reasonable accuracy is required. Since little heat is generated by a creep test, except possibly during the rapidly varying conditions which occur immediately after load application, temperature control is not difficult to achieve. Because of this more rapid variation at short times, automatic data recording is desirable, and this may also affect the value of the earliest time for which creep data is available.

B. Stress-Relaxation

Creep and stress-relaxation tests are dual techniques in which the roles of stress and strain are interchanged. For this reason the practical constraints placed on the two methods of material characterization are very similar. Since load is the measured output in a stress-relaxation test it is a simple matter to convert data into either true or nominal stress values. Due to the form of the strain input, the cross-sectional area of the specimen does not vary with time for $t > 0$ though it may be different from the value at $t = 0$.

As discussed for the creep test, practical application of an exact step-input is impossible, and must be approximated by a rapid change of strain the form of which results from a compromise between the requirements of speed and neglect of inertia. For this reason stress-relaxation data cannot be accurately measured at very short times, and there must also be a practical limit at long times. Again, the most commonly used configurations are those of extension and torsion of a rod. One method of applying the strain (3.4) is to apply a constant load (longitudinal or end torque) of a magnitude much larger than that required to produce the desired strain (even at short times) and to provide stops which prevent deformation beyond the desired amount. The stress in the specimen is then determined from load data taken from a load cell connected in series with the specimen; this load cell must be much stiffer than the specimen in order that its changing extension shall not significantly affect the strain applied to the specimen. A more elaborate method of strain application is the use of some form of servo-mechanism which reduces the applied load in such a manner as to maintain constant strain. A more tedious method is to remove loads by hand to achieve the same effect. The latter methods have the advantage of not requiring the load cell connected in series in order to measure the output.

C. Methods for Determining Complex Modulus

Dynamic methods for determining complex modulus functions were described in general terms in Sections 1C and 2 by reference to extensional deformations of a rod or strip. Similar analyses are readily given for torsional deformations of a rod and for flexural deformations of a rod, strip or beam excited either by end loads or by time-dependent loading in three- or four-point bending tests. These analyses result in similar equations, solutions and data interpretation, but involve different complex moduli and different geometrical parameters (shape factors). Forced vibrations and resonance methods have been frequently used in all of these modes of deformation and free vibration can be used in flexural and extensional modes, but is most commonly associated in the low frequency range with the torsional pendulum apparatus.

The range of frequency which can be covered by a single piece of apparatus depends on the material under consideration and upon the type of driving mechanism used. Frequencies lower than approximately 10^{-4} Hz are difficult to achieve and, if purely mechanical means of excitation are used it is difficult to achieve frequencies in excess of approximately 10^2 Hz. When electromagnetic excitation is employed it is possible to study frequency dependence up to a limit of around 10^4 Hz. In these latter cases power limitations result in very small amplitudes of vibration which in turn require electromagnetic or similar forms of data collection.

Forced vibration and resonance methods rely on similar techniques of excitation, although they employ different methods of data analysis. Due to neglect of inertia effects in the forced vibration technique, this method cannot be used at frequencies in excess of about 10^3 Hz. Analysis of experimental data for the resonance method places a practical lower limit on the frequency range for this method of approximately 10 Hz. Obviously the precise frequency range within which a particular apparatus can operate depends on material properties, specimen dimensions, added mass or inertia, and method of excitation. For standard resonance equipment the possible variation in these parameters permits a frequency range of about one decade to be covered.

Free vibration methods such as the torsional pendulum can be used to extend the frequency range below that covered by resonance methods. Again, the precise range of applicability depends on the parameters discussed above. For apparatus such as the torsional pendulum, which employs significant added mass, the frequency range is of the order 10^{-1}–10 Hz. Techniques such as the vibrating reed (see Ferry, 1961), which involve vibration of the specimen alone, extend this frequency range to approximately 10^4 Hz.

For frequencies in excess of 10^4 Hz wave propagation methods must be used. Either longitudinal or transverse waves can be propagated and, as pointed out in Section 2C, different combinations of material parameters are determined in the former case depending on whether lateral contraction can take place freely or is prevented by geometrical constraint. Longitudinal waves are most easily excited, either by detonation of a small explosive charge, by a piezoelectric transducer, or by impact of a projectile (see Kolsky, 1966). Direct excitation of transverse waves is more difficult to control, but these waves are readily generated within a material as the result of the refraction of longitudinal waves at oblique incidence to the specimen surface (see Kolsky, 1963). This method is difficult to use for highly viscoelastic materials due to rapid attenuation of the waves. Continuous waves can be generated at frequencies up to approximately 10^4 Hz, but pulse propagation techniques must be used at higher frequencies. The upper limit attained in this manner is approximately 10^7 Hz. The approximate ranges of applicability of the various methods are summarized in Fig. 4.

It will be noted that the methods discussed above permit complex modulus data to be determined within the approximate range 10^{-4}–10^7 Hz, that is, over approximately eleven decades of frequency. It should be recognized that this data must be laboriously compiled from several quite separate techniques

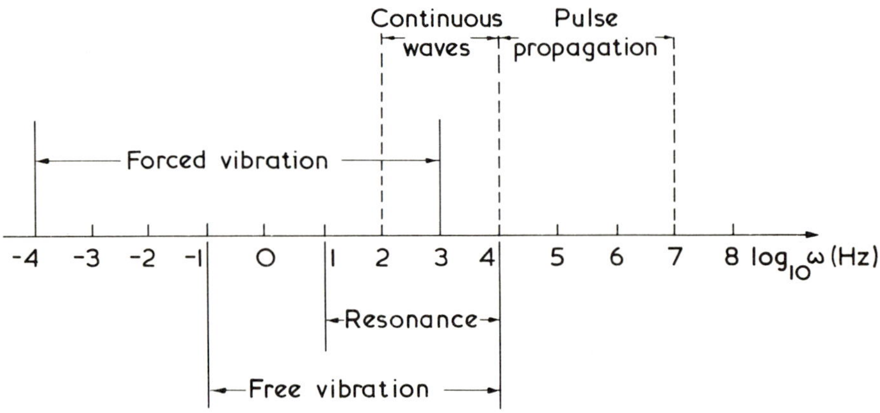

Fig. 4. Approximate frequency ranges of various forms of dynamic test.

involving different equipment. Thus the determination of material functions is not as straightforward as might have been expected from the general remarks made in Section 1C.

Free vibration and wave propagation techniques involve the measurement of quantities which are varying with time (attenuation of amplitude and passage of stress pulse) and therefore require data collection techniques which correlate data with the passage of time. Forced vibration and resonance methods both involve steady vibration situations in which amplitude and phase are the required data, and data collection is somewhat simpler for these tests. Pulse propagation involves material loading and data collection over a very short period of time during which significant temperature variation cannot occur. In this case it is only necessary to record the ambient temperature at which the data was recorded. The three vibration techniques require that the motion has continued for a sufficiently long time for transient start-up effects to have died out. For the dissipative materials discussed here this may produce a significant amount of thermo-mechanical heating, so that temperature control is more important.

4. TYPICAL MATERIAL DATA

The precise forms of the material functions depend on the material considered and also on such factors as the temperature at which the tests are performed. Temperature dependence is discussed in Section 5. Here, some typical material data is illustrated, and the relevance of spring-and-dashpot

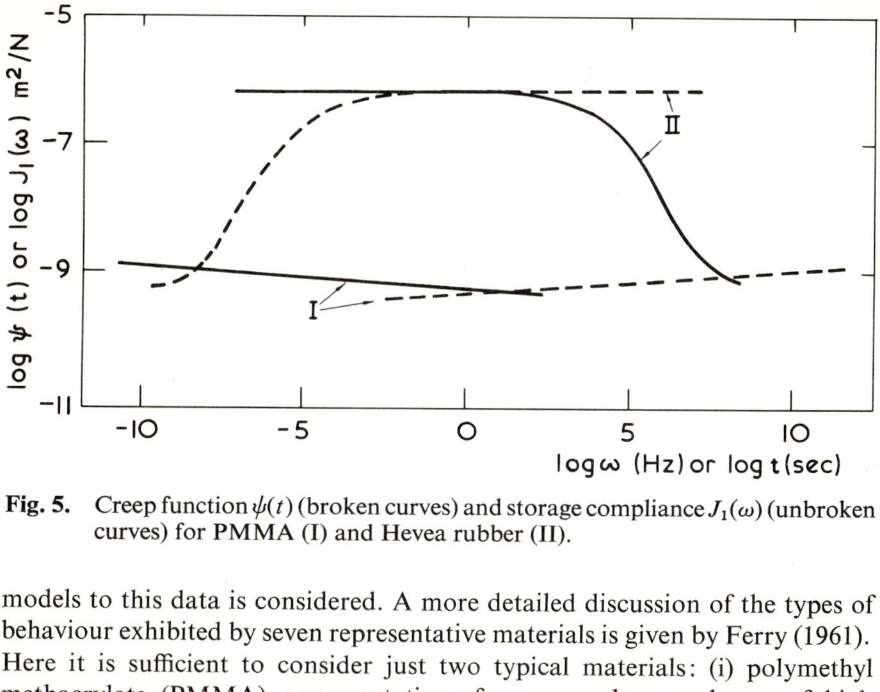

Fig. 5. Creep function $\psi(t)$ (broken curves) and storage compliance $J_1(\omega)$ (unbroken curves) for PMMA (I) and Hevea rubber (II).

models to this data is considered. A more detailed discussion of the types of behaviour exhibited by seven representative materials is given by Ferry (1961). Here it is sufficient to consider just two typical materials: (i) polymethyl methacrylate (PMMA), representative of an amorphous polymer of high molecular weight below its glass transition temperature and (ii) a lightly vulcanized Hevea rubber representative of the rubbery behaviour of a lightly crosslinked amorphous polymer. Broadly speaking the first material typifies the behaviour of other plastics such as polyethylene, ebonite, polystyrene, etc. (see Kolsky, 1960).

The sources of experimental data are referenced by Ferry (1961); the magnitude values for the two materials are not directly comparable since those for the PMMA are for shear behaviour, whereas those for the rubber are for extension. Equations (1.13) show that these moduli are related by a factor of approximately three when Poisson's ratio is approximately one half, and is insensitive to time or frequency. The two sets of data are also appropriate to different temperatures; that for the PMMA has been reduced to a reference temperature of −22°C, whereas that for the rubber has a reference temperature of 25°C. Because of the differences in material and reference temperature, the two sets of data represent typical glassy and rubbery behaviour.

The data presented are restricted to the creep and stress-relaxation functions, the real parts of the complex modulus and complex compliance, and the loss tangent. The imaginary parts of the complex functions can be calculated from

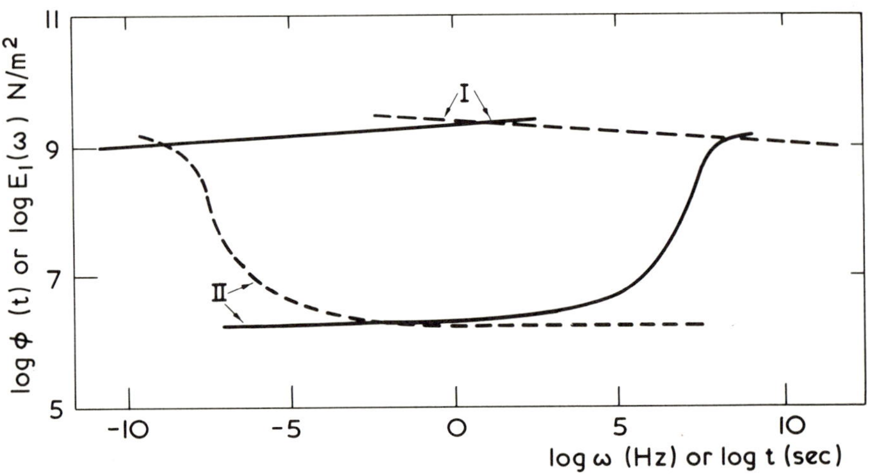

Fig. 6. Stress-relaxation function $\phi(t)$ (broken curves) and storage modulus $E_1(\omega)$ (unbroken curves) for PMMA (I) and Hevea rubber (II).

the data given, or can be found in Ferry (1961). Due to the significant variation of these material functions over the full experimental range, it is convenient to plot the curves on logarithmic axes. Fig. 5 shows values of creep function and storage compliance J_1 for both materials and Fig. 6 gives stress-relaxation and storage modulus E_1 data. These data demonstrate that the results

$$\psi(\infty) = J_1(0) = 1/E_1(0) = 1/\phi(\infty) \tag{3.31}$$

which can be obtained from (2.30), (2.33), (2.35) and (2.39) provided that $\eta = \infty$, are satisfied at the relevant ends of the experimental range. At the other ends of the experimental range the data indicate that the limiting values, if they exist, may satisfy the relations

$$\psi(0) = J_1(\infty) = 1/E_1(\infty) = 1/\phi(0). \tag{3.32}$$

For intermediate times the four curves (for each material) have forms which approximate to mirror images of each other; thus the functions

$$\psi(t), \qquad J_1(1/t), \qquad 1/E_1(1/t), \qquad 1/\phi(t) \tag{3.33}$$

are approximately equal. However, it can be shown that

$$\psi(t)\,\phi(t) \leqslant 1, \qquad E_1(1/t) > \phi(t), \qquad J_1(1/t) < \psi(t) \tag{3.34}$$

at all times.

Variation of the loss tangent with frequency is shown in Fig. 7 for the two typical materials. The curve for PMMA shows little dependence on frequency whereas that for the rubber varies by two orders of magnitude having a loss

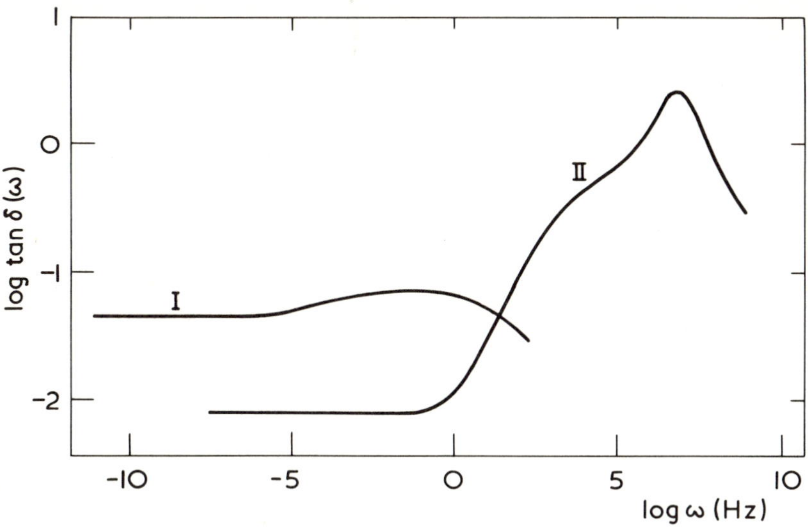

Fig. 7. Loss tangent for PMMA (I) and Hevea rubber (II).

peak at approximately 10^5 Hz. In principle a material may exhibit a number of such loss peaks, each corresponding to some molecular mechanism.

Using the results of Section 4 (Chapter 2) it is readily seen that the loss tangents for the simple spring-and-dashpot models are

Maxwell: $E/\omega\eta$

Voigt: $\eta\omega/E$

Standard linear solid: $\eta E_2\,\omega[E_1(E_1 + E_2) + \eta^2\,\omega^2]^{-1}$. (3.35)

At low frequencies the loss tangent for a standard linear solid has a frequency dependence similar to that for a Voigt model: at higher frequency it is similar to that for a Maxwell model. The frequency dependence of these models is shown in Fig. 8 for a particular choice of model parameters. Variation of these parameters causes translation of the curves parallel to the coordinate axes but no change of shape.

It is evident from comparison of Figs. 7 and 8 that none of the simple spring-and-dashpot models can model the behaviour of PMMA or the rubber over the whole frequency range. However, it may be adequate to model the rubber by a Voigt model in the range $1 < \omega < 10^5$ Hz, by a Maxwell model in the range $10^5 < \omega < 10^9$ Hz, or by a standard linear solid over the range

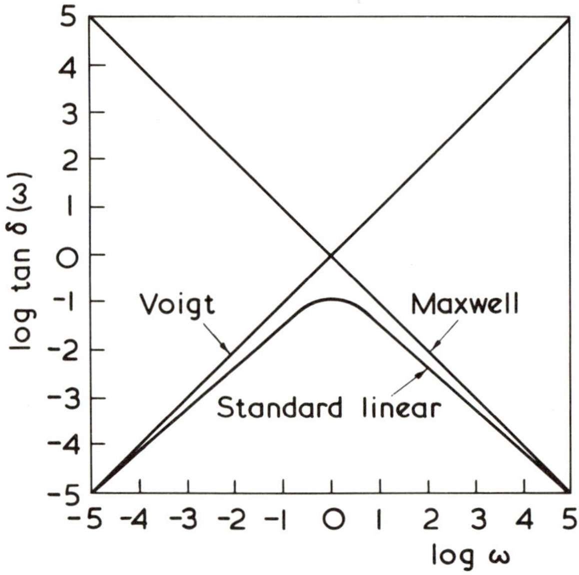

Fig. 8. Variation of loss tangent with frequency for the simple spring-and-dashpot models.

$1 < \omega < 10^9$ Hz. For applications involving frequencies within only one of these ranges the appropriate spring-and-dashpot model may be adequate. The loss tangent of the rubber in the range $10^{-7} < \omega < 1$ Hz, and of PMMA over the entire range of data, is approximately independent of frequency, and cannot be modelled by a simple combination of springs and dashpots. In this case, however, the approximation preceding (2.40) is valid, and the simple analytic form given in that equation applies; this is confirmed by the linear variation of $\log E_1(\omega)$ with $\log \omega$ over the appropriate frequency ranges in Fig. 6.

5. VARIATION OF MATERIAL PROPERTIES WITH TEMPERATURE

The typical material data discussed in the previous section is appropriate to particular reference temperatures and it is pertinent to enquire how this data would have differed if other reference temperatures had been chosen. For the majority of polymeric materials the various curves plotted in Figs. 5–7 do not show a radical change in shape with change in temperature, but they tend to suffer a translation parallel to the $\log \omega$ or $\log t$ axes. With an increase in temperature the creep or stress-relaxation processes are accelerated and these property functions are translated in the negative direction on the $\log t$ axis. Due to the reciprocal nature of time and frequency, an increase in temperature causes a movement of the complex material functions to higher frequencies.

For a significant number of materials the shift of property functions parallel to the $\log t$ or $\log \omega$ axes without change of shape is sufficiently accurate for this fact to be used as an aid in property measurement and to form the basis of a thermo-viscoelastic theory. The assumed behaviour has been termed thermo-rheologically simple by Schwarzl and Staverman (1952) and is often referred to as time-temperature equivalence. An estimate for the amount of shift required for a given temperature change is provided by the Williams-Landel-Ferry (WLF) equation (see Ferry, 1961). The validity of time-temperature equivalence has been demonstrated most completely for polyisobutylene. Fitzgerald *et al.* (1953) obtained data in the frequency range $10 < \omega < 10^4$ Hz for the shear compliance at 22 different temperatures between −45°C and 100°C. Their data, which is shown in part in Fig. 9(a), was then used with the WLF equation to construct the master curve shown in Fig. 9(b) which covers nine decades of frequency at a single temperature. The scatter of the experimental data about this curve is very slight, indicating that time-temperature equivalence is a valid approximation for the material investigated.

When there is reason to suppose that a material does behave in the manner discussed above, experimentation at different temperatures can be used to extend the data range beyond that for which direct methods of measurement are available. Such an extension of the data range is likely to be valid if the range over which data is actually measured is sufficiently large and if time-temperature equivalence is demonstrated to be valid within that range. This accounts for data being quoted, for example in Figs. 5–7, for times or frequencies which are not directly attainable using existing experimental techniques.

Since thermo-rheologically-simple materials must have their properties evaluated on a time scale which is modified by temperature, it is not difficult to incorporate their behaviour into a theory of thermo-viscoelasticity. Morland and Lee (1960) have shown that the constitutive relation (2.5) is replaced by

$$\sigma(t) = \int_{-\infty}^{t} \phi(\xi - \xi')\, \dot{\epsilon}(\tau)\, d\tau \tag{3.36}$$

where the dot denotes the material time derivative with respect to τ, and

$$\xi(t) = \int_{0}^{t} a[T(t')]\, dt', \qquad \xi = \xi(t), \qquad \xi' = \xi(\tau) \tag{3.37}$$

where the material is supposed to be at a constant reference temperature T_0 for $t < 0$. In these equations $\phi(t)$ is the stress-relaxation function at the refer-

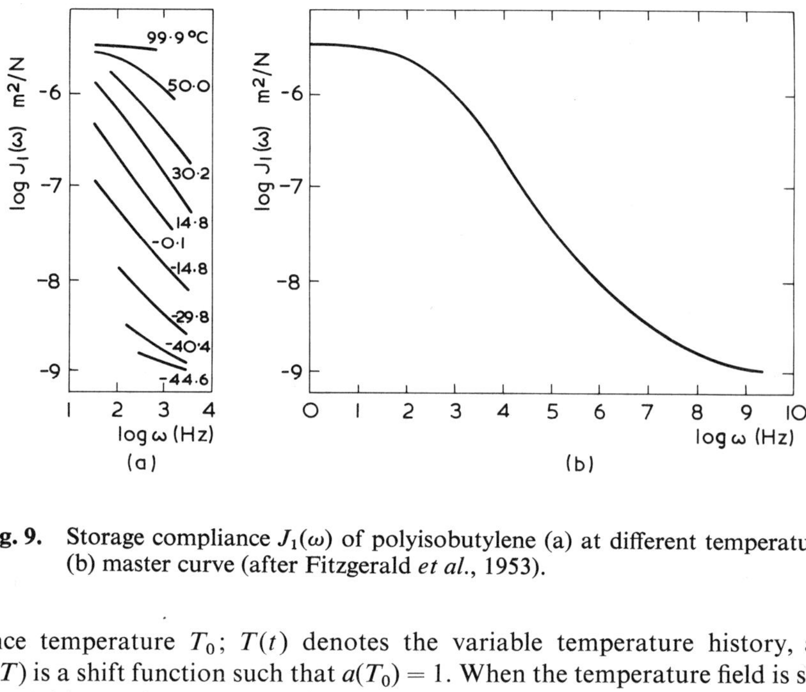

Fig. 9. Storage compliance $J_1(\omega)$ of polyisobutylene (a) at different temperatures, (b) master curve (after Fitzgerald *et al.*, 1953).

ence temperature T_0; $T(t)$ denotes the variable temperature history, and $a(T)$ is a shift function such that $a(T_0) = 1$. When the temperature field is such that $T(t) \equiv T_0$ for $t \geqslant 0$, equation (3.37) shows that the reduced time ξ and real time t are identical, and (3.36) reduces to (2.5). When $T(t) \equiv T_1 \neq T_0$ for $t \geqslant 0$, (3.37) yields

$$\xi(t) = a(T_1)\, t, \qquad \xi' = a(T_1)\, \tau \tag{3.38}$$

and the stress-relaxation function in (3.36) suffers a shift along the logarithmic time axis given by

$$\log \xi = \log a(T_1) + \log t. \tag{3.39}$$

When the temperature field $T(t)$ does not assume a constant value for $t \geqslant 0$, the reduced time ξ is given by (3.37) which takes account of the entire past history of the temperature field. For derivation and justification of (3.36) and (3.37) the reader is referred to the original paper by Morland and Lee (1960).

Finally it is noted that the condition of time-temperature equivalence is not a universally valid rule. In order to increase its range of applicability, some attempts have been made to model data by including a magnification factor (that is a shift of the curves parallel to the logarithmic modulus axis) in addition to the shift along the logarithmic time or frequency axis. Such devices may prove to be useful, but for some materials temperature dependence may not follow a simple law, and the interpretation of data and its use in solving boundary value problems, is then much more difficult.

6. ANISOTROPIC BEHAVIOUR

The constitutive relation (2.44) for an isotropic material involves two material functions, and complete determination of material behaviour is achieved through separate experiments involving two different modes of deformation. When the material behaviour is anisotropic a larger number of material functions are required to characterize the material, and it follows that a corresponding number of different modes needs to be considered in the experimental determination. In this context simple extension in three mutually orthogonal directions provides three different modes for the most general form of anisotropy, and shear deformations in the three coordinate planes provide a further three. The most appropriate tests for a given material will depend on the nature of the particular form of anisotropy exhibited by the material.

It was noted in Section 3C of Chapter 2 that the most general form of anisotropy requires 36 material functions to describe the behaviour of a linear viscoelastic material. In principle these functions could be determined from six experiments in each of which only one input function (stress or strain) was non-zero. In each experiment all six output functions (strain or stress) would have to be measured. Each of the 36 sets of data would then be analysed, in the context of a one-dimensional relation, using the relevant technique from those discussed in earlier sections of this chapter. For this general form of anisotropy it is evident that this would require a considerable experimental programme to determine the material functions over even a modest range of frequency or time.

When the material possesses some less general form of anisotropy the number of material functions, and therefore also the size of the experimental programme, is reduced. As discussed in Chapter 2, Section 3C this results in a larger number of functions than is required for an elastic material having the same material symmetry, since the assumption $c_{ijkl} = c_{klij}$ is not valid in general. In some cases it may be that experimental data will confirm the validity of this relation (see, for example, Halpin and Pagano, 1968), but it cannot be assumed *a priori*.

It is likely that experimental data for the viscoelastic behaviour of anisotropic materials will be forthcoming only for special simple forms. The experimental data of Halpin and Pagano (1968) referred to above were obtained for plane stress conditions in a sheet of material which was deformed by loads acting in its own plane. Since these data characterize the relationship between three stress components and three strain components, the problem is reduced to determination of nine functions at most. The material considered was a unidirectional fibre reinforced material consisting of nylon fibres in a rubber matrix. Creep tests were performed with unidirectional loading parallel and transverse to the fibre direction and, in lieu of a shear test, unidirectional loading at an intermediate angle (30°) to the fibre direction. In each test three in-plane strains were measured and the data indicated that the symmetry $c_{ijkl} = c_{klij}$ was valid, within the limits of experimental accuracy, and for the material and restricted geometry considered. For more general situations, and by analogy with the special case of plane stress discussed above, it may be convenient to employ tests in which more than one input component is non-zero. A possible programme of tests, which does not assume that $c_{ijkl} = c_{klij}$ and therefore provides an experimental test of the validity of this relation, has been given by Hayes and Morland (1969). For particular simple forms of anisotropy an experimental programme can also be modelled on the corresponding requirements for elastic materials, but must be supplemented to provide extra tests to check the symmetry condition.

Chapter 4 Nonlinear Theory of Viscoelasticity

In this chapter an outline is given of the main features of theories proposed to describe nonlinear mechanical behaviour of viscoelastic materials. The mathematical development is purely formal, attempting to give the reader whose main concern is not one of mathematical detail some idea of the steps through which the formulation of the theories passes with a statement of the main assumptions upon which they are based. The reader who is interested in rigorous mathematical details can consult one of the books devoted to that approach (see, for example, Eringen, 1962; Truesdell and Noll, 1965). No attempt is made here to mention all the proposed theories. The discussion is restricted to those which have either been applied successfully to the interpretation of experimental data or, in my opinion, appear to be suitable starting points for future work in this field.

1. FUNDAMENTALS OF THE THEORY

A. Characterization with Respect to a Natural Reference State

Consider a body of material in an initial virgin state of zero stress and zero strain, and denote by (X_1, X_2, X_3) the coordinates of a typical material particle relative to a fixed Cartesian coordinate system. At some later time τ, when the material has undergone deformation, the coordinates of the typical particle are denoted by

$$x_i(X_j, \tau), \qquad (i,j = 1, 2, 3) \tag{4.1}$$

that is, they depend on which material particle is being discussed and on the particular point in time. The assumption of an initial natural reference state X_i is most appropriate when the material has a preferred configuration; this is true for many solid-like materials but may be less appropriate for fluid-like behaviour where the material may occupy more than one stress-free and strain-free configuration. Although the present approach is valid for both types of material behaviour, it is most appropriate to solids, with which this book is mainly concerned. An alternative approach will be considered in Section 3.

In the nonlinear theory it cannot be assumed *a priori* that the classical strain measure $\boldsymbol{\epsilon}$ is an adequate and valid measure of deformation. First consider a one-dimensional situation in which $x = x(X, \tau)$. A measure of the deformation in the material at a given time and at a given point in the material is the relative displacement of two neighbouring material points, that is

$$x(X+\delta X,\tau)-x(X,\tau)=\delta X\frac{\partial x(X,\tau)}{\partial X}+\tfrac{1}{2}(\delta X)^2\frac{\partial^2 x(X,\tau)}{\partial X^2}+\cdots \tag{4.2}$$

In this one-dimensional case, the deformation measure can therefore be expressed as some function of the derivatives

$$\frac{\partial x}{\partial X},\qquad \frac{\partial^2 x}{\partial X^2},\cdots \tag{4.3}$$

For three-dimensional behaviour the deformation measure is generalized to depend on

$$F_{rs}(X_i,\tau)=\frac{\partial x_r}{\partial X_s},\qquad F_{rs}^{(n)}=\frac{\partial^n x_r}{\partial X_s^n},\qquad G_{rmn}=\frac{\partial^2 x_r}{\partial X_m\,\partial X_n},\ \text{etc.} \tag{4.4}$$

and a viscoelastic material may be defined as one for which stress depends on the entire history of all of the quantities (4.4). In particular, a "simple material" is defined as one for which stress depends only on the history of the deformation gradients F_{rs}. This book is restricted to a discussion of the behaviour of simple materials. Theories for materials having dependence on higher deformation gradients, which may be relevant to materials with macro-structure, have been studied by many theoretical workers (see, for example, Eringen and Suhubi, 1964; Green and Rivlin, 1964, 1965; Green *et al.*, 1965) but they have had little practical impact.

From the physical definition of a simple material it follows that

$$\sigma_{ij}(t)=\mathop{\mathscr{F}_{ij}}_{\tau=-\infty}^{t}\{F_{rs}(\tau)\} \tag{4.5}$$

where dependence of σ_{ij} and F_{rs} on the spatial variables X_j has not been written explicitly. Since attention is restricted to homogeneous materials, the functionals $\mathscr{F}_{ij}$ do not depend explicitly on $\mathbf{X}$. It will be seen below that these functionals cannot be given arbitrary forms but are subject to certain restrictions which all materials must satisfy. For simplicity (4.5) is written in matrix notation

$$\boldsymbol{\sigma}(t)=\mathop{\mathscr{F}}_{\tau=-\infty}^{t}\{\mathbf{F}(\tau)\} \tag{4.6}$$

and it is noted that the polar decomposition theorem (see Truesdell and Noll, 1965) allows $\mathbf{F}$ to be written in either of the unique forms

$$\mathbf{F}=\mathbf{RU}=\mathbf{VR} \tag{4.7}$$

where **R** is an orthogonal matrix and **U** and **V** denote positive definite symmetric matrices. The physical interpretation of (4.7) is that the deformation represented by **F** can be considered as a rigid rotation **R** which is either followed by a pure stretch **V** or preceded by a pure stretch **U**.

A restriction on the form of functional appearing in (4.6) arises through the principle of "objectivity" or "material frame indifference". Essentially this principle is a statement of the fact that the mechanical properties of a material depend only on the material itself and not on the observer. Of course, certain numerical values of parameters do depend on the coordinate system chosen, but the type of material behaviour is frame-independent. When the consequences of this restriction are applied to the constitutive equation (4.6) it can be shown (see Truesdell and Noll, 1965) that the equation must be expressible in the form

$$\boldsymbol{\sigma}(t) = \mathbf{F}(t) \underset{\tau=-\infty}{\overset{t}{\mathbf{G}_1}} \{\mathbf{U}(\tau)\}\, \mathbf{F}^T(t) \tag{4.8}$$

where $\mathbf{F}^T$ is the transpose of **F**, and $\mathbf{G}_1$ is a new arbitrary functional.

An alternative form to (4.8) is obtained by substituting from (4.7) into (4.8) so that

$$\boldsymbol{\sigma}(t) = \mathbf{R}(t) \underset{\tau=-\infty}{\overset{t}{\mathbf{G}}} \{\mathbf{U}(\tau)\}\, \mathbf{R}^T(t) \tag{4.9}$$

and explicit dependence on $\mathbf{U}(t)$ has been included in the new arbitrary functional **G**. Expression (4.9) shows that the principle of objectivity implies that current stress depends only on the current value of rotation **R**, but depends on the entire past history of the stretch **U**.

Since **R** is an orthogonal matrix, $\mathbf{R}^T\mathbf{R} = \mathbf{R}\mathbf{R}^T = \mathbf{I}$ and (4.9) gives

$$\boldsymbol{\Sigma}(t) \equiv \mathbf{R}^T \boldsymbol{\sigma} \mathbf{R} = \underset{\tau=-\infty}{\overset{t}{\mathbf{G}}} \{\mathbf{U}(\tau)\} \tag{4.10}$$

where $\boldsymbol{\Sigma}$ is the stress matrix evaluated in a coordinate system which is rotated with the rigid-body-rotation component of the deformation. The principle of objectivity shows, through (4.10), that a permissible form of constitutive relation is one in which $\boldsymbol{\Sigma}$ is expressed as an arbitrary functional of the deformation measure **U**.

In the terminology of the mathematical theory, **U** is an objective strain measure. However, it is not the only such measure since any simple function of **U** (for example $\mathbf{U}^{1/2}$, $\log \mathbf{U}$, etc.) would also satisfy objectivity requirements. In fact **U** is not a convenient measure since it requires evaluation of the decomposition (4.7).

The right Cauchy-Green strain measure **C** defined by

$$\mathbf{C} = \mathbf{F}^T\mathbf{F} = \mathbf{U}^T\mathbf{R}^T\mathbf{R}\mathbf{U} = \mathbf{U}^T\mathbf{U} = \mathbf{U}^2 \tag{4.11}$$

is objective since it is a simple function of **U**, and it has the advantage that it is readily calculated from the first part of (4.11). A slightly different strain measure **E**, which will be used extensively in later sections, is defined by

$$2\mathbf{E} = \mathbf{C} - \mathbf{I} = \mathbf{F}^T\mathbf{F} - \mathbf{I} \tag{4.12}$$

which is also objective. The reason for introducing this measure can be seen by first considering the displacement vector **u** which has components

$$u_i = x_i - X_i. \tag{4.13}$$

Then, noting that $F_{ij} = \partial x_i/\partial X_j$, (4.12) gives

$$2E_{ij} = \frac{\partial x_k}{\partial X_i}\frac{\partial x_k}{\partial X_j} - \delta_{ij} = \left[\frac{\partial u_k}{\partial X_i} + \delta_{ik}\right]\left[\frac{\partial u_k}{\partial X_j} + \delta_{jk}\right] - \delta_{ij}$$

$$= \frac{\partial u_i}{dX_j} + \frac{\partial u_j}{\partial X_i} + \frac{\partial u_k}{dX_i}\frac{\partial u_k}{dX_j}$$

which can be written in the form

$$E_{ij} = \epsilon_{ij} + \frac{1}{2}\frac{\partial u_k}{\partial X_i}\frac{\partial u_k}{\partial X_j}. \tag{4.14}$$

Thus, when the displacement gradients are small, **E** reduces to the classical strain $\boldsymbol{\epsilon}$. In general **E** and $\boldsymbol{\epsilon}$ are not directly related, and $\boldsymbol{\epsilon}$ is not an objective strain measure. Thus the classical linear theories of elasticity and viscoelasticity are not frame-indifferent; they are approximations which are valid only under certain conditions (see Section 2B).

Because of the simple relations connecting **C**, **E** and **U** the constitutive relation (4.10) may be written in the alternative forms

$$\boldsymbol{\Sigma}(t) = \mathop{\mathscr{H}_1}_{\tau=-\infty}^{t} \{\mathbf{C}(\tau)\} \tag{4.15}$$

$$\boldsymbol{\Sigma}(t) = \mathop{\mathscr{H}_2}_{\tau=-\infty}^{t} \{\mathbf{E}(\tau)\} \tag{4.16}$$

where $\mathscr{H}_1$ and $\mathscr{H}_2$ are new functionals, which are arbitrary with respect to objectivity requirements. Their detailed forms have to be determined by a combination of theory and experiment for any particular material which will be discussed in detail in later chapters.

B. Characterization in the Absence of a Natural Reference State

As remarked earlier, it may be inappropriate in some instances to formulate the theory with respect to a reference state **X** in which the material is in a state of zero stress and zero strain. For an elastic solid this corresponds to the unique natural state of the material. However, for a viscous fluid there are an infinite number of configurations which are possible natural states in the sense that they can exist without the imposition of external forces; in such a case the choice of one such configuration as a reference state is bound to be arbitrary. When a viscoelastic material has a well-defined reference configuration it may be appropriate to use the formulation of the theory given above. However, when no such configuration is evident it is preferable to use an alternative formulation rather than introduce some arbitrariness.

In the absence of a *natural* reference state it is possible to use the *current* configuration as the reference state and define past history with respect to it. If $x_j(t)$ denote the current position of a particle, then the position-history of the particle can be expressed as

$$x_i(\tau) = x_i[x_j(t), \tau] \tag{4.17}$$

that is, $x_i(\tau)$ denotes the location at time τ of the particle that will occupy the position $x_j(t)$ at time t. By analogy with the deformation gradient $\mathbf{F}$, a *relative* deformation gradient $\mathbf{F}_t$ is defined by

$$\mathbf{F}_t = \left\| \frac{\partial x_i(\tau)}{\partial x_j(t)} \right\| \tag{4.18}$$

together with other relative matrices, for example, $\mathbf{C}_t = \mathbf{F}_t^T \mathbf{F}_t$.

If so desired, all material constitutive relations could be expressed in terms of relative strain measures, whether the material had a natural reference state or not. One such form of constitutive relation for a simple material is (see Truesdell and Noll, 1965)

$$\mathbf{R}^T \boldsymbol{\sigma} \mathbf{R} = \mathop{\mathscr{F}}_{\tau=-\infty}^{t} \{\mathbf{R}^T(t)\, \mathbf{C}_t(\tau)\, \mathbf{R}(t); \mathbf{C}(t)\}. \tag{4.19}$$

If the material has no natural reference state, dependence on $\mathbf{C}(t)$ must be omitted from (4.19). However, since (4.19) contains the non-relative matrix $\mathbf{R}$ in a more complex manner, it is apparent that (4.19) cannot represent a material with no natural reference state without some modification and this is discussed next.

C. Isotropic Materials

When a nonlinear material is deformed its state of anisotropy, as defined by its response to further loads, changes as the deformation proceeds. An isotropic material can be defined as a material which is isotropic with respect to loads applied to a specimen which has had no previous loading history. For a material with a natural reference state, this places a restriction on the material behaviour when it is loaded in that reference state. In the case of the material represented by (4.16), no simplification in the form of that equation results, but the functional $\mathscr{H}$ becomes an isotropic functional (see Truesdell and Noll, 1965). An explicit form for $\mathscr{H}$ will be considered in Section 2A.

When the material has no natural reference state the assumption of isotropy must apply to all possible virgin states. When this assumption is applied to the constitutive relation (4.19) there results the simpler form

$$\boldsymbol{\sigma}(t) = \mathop{\mathbf{G}}_{\tau=-\infty}^{t} \{\mathbf{C}_t(\tau); \mathbf{B}(t)\} \tag{4.20}$$

where $\mathbf{G}$ is an isotropic functional. In this relation $\mathbf{B}$ denotes the left Cauchy-Green strain measure defined by

$$\mathbf{B} = \mathbf{F}\mathbf{F}^T = \mathbf{V}\mathbf{R}\mathbf{R}^T\mathbf{V}^T = \mathbf{V}^2. \tag{4.21}$$

It is noted that the assumption of isotropy has the effect of removing the dependence on $\mathbf{R}(t)$ from (4.19). The possibility of a natural reference state is retained explicitly in (4.20) through the dependence on $\mathbf{B}(t)$, and implicitly through $\mathbf{C}_t(-\infty)$ if it is assumed that the material occupied the reference configuration at $\tau = -\infty$. When no natural state exists the dependence on $\mathbf{B}(t)$ can be omitted from (4.20) and there results

$$\boldsymbol{\sigma}(t) = \mathop{\mathbf{G}}_{\tau=-\infty}^{t} \{\mathbf{C}_t(\tau); \rho\} \tag{4.22}$$

where ρ denotes density, which is the constitutive relation for a simple fluid.

The assumption of isotropy is responsible for reducing (4.19) to the form (4.20), and the form (4.22) for a material with no natural reference state is readily determined. It has been shown (see Truesdell and Noll, 1965) that simple materials with no reference state are necessarily isotropic, so that (4.22) is the most general material of that form.

2. INTEGRAL CONSTITUTIVE RELATIONS

The general constitutive relations presented in the previous section are formal expressions which would have little practical value for modelling the properties of real materials. In order to obtain relations with potential practical value, it is necessary that more explicit expressions should be proposed for representing the functional dependences. In the same way that it was possible to express the classical linear constitutive relation in a number of alternative forms (see Chapter 2), it is also possible to express the nonlinear relation in an integral, differential or complex form. These formulations are presented in the present section and in Sections 3 and 5 respectively.

A. Green-Rivlin Theory

The present discussion is based on the form (4.16) of constitutive relation for a material with a natural reference state, that is,

$$\boldsymbol{\Sigma}(t) \equiv \mathbf{R}^T \boldsymbol{\sigma} \mathbf{R} = \underset{\tau=-\infty}{\overset{t}{\mathscr{H}}} \{\mathbf{E}(\tau)\} \tag{4.23}$$

where $\mathscr{H}$ is supposed to be an isotropic functional, so that (4.23) is the constitutive relation for an isotropic material. It is assumed that $\mathscr{H}$ is expressible as an integral polynomial; the mathematical implications of this assumption will not be considered here since they are not readily interpreted in physical terms. Compared with the assumptions upon which the theory is based these limitations are probably not crucial, and the final assessment of the theory can come only from its ability to model experimental data and predict real behaviour.

In the same spirit it is supposed that (4.23) can be inverted to yield an equation of the form

$$\mathbf{E}(t) = \underset{\tau=-\infty}{\overset{t}{\mathbf{S}}} \{\boldsymbol{\Sigma}(\tau)\} \tag{4.24}$$

expressing current strain as a functional of stress history.

In the discussion that follows it is convenient to consider (4.23) and (4.24) at the same time by writing

$$\mathbf{Q}(t) = \underset{\tau=-\infty}{\overset{t}{\mathscr{F}}} \{\mathbf{P}(\tau)\} \tag{4.25}$$

where $\mathbf{Q}$ and $\mathbf{P}$ can be interpreted either as

$$\mathbf{Q} = \boldsymbol{\Sigma} = \mathbf{R}^T \boldsymbol{\sigma} \mathbf{R}, \qquad \mathbf{P} = \mathbf{E} \tag{4.26}$$

or

$$\mathbf{Q} = \mathbf{E}, \qquad \mathbf{P} = \boldsymbol{\Sigma} = \mathbf{R}^T \boldsymbol{\sigma} \mathbf{R}. \tag{4.27}$$

For the homogeneous isotropic materials considered here the functional $\mathscr{F}$ is an isotropic functional: a polynomial integral expression for $\mathscr{F}$ has been given by Green and Rivlin (1957, 1960) and Green *et al.* (1959). A form which is convenient for present purposes is (Lockett, 1965)

$$\begin{aligned}\mathbf{Q}(t) = &\int_{-\infty}^{t} \{\mathbf{I}\psi_1 T_1 + \psi_2 \mathbf{M}_1\}\, d\tau_1 + \iint_{-\infty}^{t} \{\mathbf{I}\psi_3 T_1 T_2 + \mathbf{I}\psi_4 T_{12} + \psi_5 T_1 \mathbf{M}_2 \\ &+ \psi_6 \mathbf{M}_1 \mathbf{M}_2\}\, d\tau_{(2)} + \iiint_{-\infty} \{\mathbf{I}\psi_7 T_{123} + \mathbf{I}\psi_8 T_1 T_{23} + \psi_9 T_1 T_2 \mathbf{M}_3 \\ &+ \psi_{10} T_{12} \mathbf{M}_3 + \psi_{11} T_1 \mathbf{M}_2 \mathbf{M}_3 + \psi_{12} \mathbf{M}_1 \mathbf{M}_2 \mathbf{M}_3\}\, d\tau_{(3)} + \cdots\end{aligned} \tag{4.28}$$

where $d\tau_{(n)} = d\tau_1 \, d\tau_2 \ldots d\tau_n$ and

$$\mathbf{M}_\alpha = \dot{\mathbf{P}}(\tau_\alpha), \qquad T_\alpha = \operatorname{tr}\mathbf{M}_\alpha,$$
$$T_{\alpha\beta} = \operatorname{tr}\mathbf{M}_\alpha \mathbf{M}_\beta, \qquad T_{\alpha\beta\gamma} = \operatorname{tr}\mathbf{M}_\alpha \mathbf{M}_\beta \mathbf{M}_\gamma. \tag{4.29}$$

When ageing of the material is included in the theory the material functions ψ_α are functions of t and of the integration variables τ_α. For a hereditary or non-ageing material ψ_1 and ψ_2 are functions of $t-\tau_1$; $\psi_3, \ldots, \psi_6$ are functions of $t-\tau_1$, $t-\tau_2$; $\psi_7, \ldots, \psi_{12}$ are functions of $t-\tau_1$, $t-\tau_2$ and $t-\tau_3$. Depending on the choice of definitions (4.26) and (4.27) the ψ_α are stress-relaxation or creep functions of various orders; an nth order function is a function of n variables. The polynomial expression (4.28) may be continued to higher orders, but no practical application has been made of such an extended relation. The terms written explicitly in (4.28) involve twelve material functions, and it will be seen in Chapter 6 that their complete experimental determination is a prohibitive task.

It is readily shown (Lockett, 1965) that the material functions ψ_α can be assumed, without loss in generality, to have certain symmetries with respect to interchange of their arguments:

$$\psi_3 = S(1,2) \qquad \psi_4 = S(1,2) \qquad \psi_6 = S(1,2)$$
$$\psi_7 = S(1,2,3) \qquad \psi_8 = S(2,3) \qquad \psi_9 = S(1,2)$$
$$\psi_{10} = S(1,2) \qquad \psi_{11} = S(2,3) \qquad \psi_{12} = S(1,3) \tag{4.30}$$

where $S(i,j)$ denotes symmetry with respect to interchange of the ith and jth arguments.

B. Special Linear Forms

A number of special forms of the Green-Rivlin theory are considered in detail in later chapters: one-dimensional behaviour in Chapter 5, incompressible materials and conditions of plane stress and plane strain in Chapter 6. Derivation of the relevant constitutive relations is presented in the chapters concerned. In the present sub-section a number of linearized forms are noted.

If the strain $\mathbf{E}$ is small in some sense (it tends to zero as displacement gradients become smaller), a first approximation to (4.28) is obtained by retaining only first order terms in that equation. With the interpretation (4.26) this yields

$$\boldsymbol{\sigma}(t) = \mathbf{R}(t) \int_{-\infty}^{t} \{\mathbf{I}\psi_1 \operatorname{tr}\dot{\mathbf{E}}_1 + \psi_2 \dot{\mathbf{E}}_1\} \, d\tau_1 \, \mathbf{R}^T(t). \tag{4.31}$$

If, in addition, rotation of material elements is assumed to be small, $\mathbf{R} \simeq \mathbf{I}$ but $\mathbf{E}$ is unaffected (since $2\mathbf{E} = \mathbf{U}^2 - \mathbf{I}$), and (4.31) becomes

$$\boldsymbol{\sigma}(t) = \int_{-\infty}^{t} \{\mathbf{I}\psi_1 \operatorname{tr} \dot{\mathbf{E}}(\tau) + \psi_2 \dot{\mathbf{E}}(\tau)\}\, d\tau. \tag{4.32}$$

Finally, if it is further assumed that displacement gradients $\partial u_i/\partial X_j$ are small, (4.14) shows that $\mathbf{E} \simeq \boldsymbol{\epsilon}$, and (4.32) becomes

$$\boldsymbol{\sigma}(t) = \int_{-\infty}^{t} \{\mathbf{I}\psi_1 \operatorname{tr} \dot{\boldsymbol{\epsilon}}(\tau) + \psi_2 \dot{\boldsymbol{\epsilon}}(\tau)\}\, d\tau \tag{4.33}$$

which is equivalent to (2.44) with $\psi_1 = \lambda$, $\psi_2 = 2\mu$; this justifies the relation (2.44) used in the classical linear theory. Any of the special linear forms (4.31), (4.32) or (4.33) may be used in situations where the relevant assumptions are justified.

C. Other Integral Relations

The constitutive relation discussed in Section 2A is derived by assuming that the functional relation (4.23) can be expressed as an integral polynomial. Several forms for the constitutive relation are given in Section 1, and each can be used as the basis of an alternative integral polynomial relation. In fact the theory originally proposed by Green and Rivlin (1957) was based on the functional relation

$$\boldsymbol{\sigma}(t) = \mathbf{F}(t) \underset{\tau=-\infty}{\overset{t}{\mathscr{F}^*}} \{\mathbf{C}(\tau)\}\, \mathbf{F}^T(t) \tag{4.34}$$

which is readily derived from (4.8) and (4.11). From a practical point of view, an integral polynomial expression based on (4.23) is preferable to one based on (4.34), since $\mathbf{E} \to \mathbf{0}$ but $\mathbf{C} \to \mathbf{I}$ as the deformations become small. In addition it will be seen from the discussion of Chapter 6 that the rotated stress $\boldsymbol{\Sigma}$ is more convenient for the interpretation of experimental data than would be the convected stress $\mathbf{F}^T \boldsymbol{\sigma} \mathbf{F}$ required by (4.34).

Of the other possible proposed functional relations, the form which has been most widely used in an integral form to model experimental data is the constitutive functional (4.20). This equation is the basis for Coleman and Noll's (1961) theory of finite linear viscoelasticity. This theory, with a description of methods for the experimental determination of material functions, is given in Chapter 7. Other forms of integral constitutive relation are presented in the theoretical literature, but they will not be discussed here.

3. DIFFERENTIAL FORMULATION

Formulations of the nonlinear theory will now be considered which generalize the linear differential formulation of (2.16). First it is necessary to introduce quantities which denote time derivatives of stress and strain. By an extension of the arguments presented in Section 1A these derivatives should be frame-indifferent quantities, and it has been shown that partial derivatives and material derivatives do not satisfy this requirement (see Oldroyd, 1950; Rivlin and Ericksen, 1955; Truesdell and Noll, 1965). Frame-indifferent time derivatives of the strain measure can be written in the form of a set of kinematic matrices (Rivlin and Ericksen, 1955):

$$\mathbf{A}_1 = \left\| \frac{\partial v_i}{\partial x_j} + \frac{\partial v_j}{\partial x_i} \right\| \tag{4.35}$$

$$\mathbf{A}_{n+1} = \frac{D\mathbf{A}_n}{Dt} + \mathbf{A}_n\,\mathbf{V} + (\mathbf{A}_n\mathbf{V})^T, \qquad n = 1, 2, \ldots \tag{4.36}$$

where v_i is the velocity field $v_i = Du_i/Dt$, D/Dt denotes the material time derivative (1.7), and

$$\mathbf{V} = \|\partial v_i/\partial x_j\|\,. \tag{4.37}$$

Since the differential formulation is often applied to describe the behaviour of viscoelastic fluids v_i is expressed as a function of $x_i(t)$ and τ, and the kinematic matrices are of relative type [cf. equation (4.18)].

When velocities and velocity gradients are small, $D/Dt \simeq \partial/\partial t$ and it follows from (4.35) and (4.36) that

$$\mathbf{A}_1 = 2\mathbf{d}, \qquad \mathbf{A}_{n+1} = 2\partial^n \mathbf{d}/\partial t^n \tag{4.38}$$

where $\mathbf{d}$ is the classical strain rate matrix (1.15).

Frame-indifferent forms of time derivative of the stress matrix are denoted by $\overset{(n)}{\boldsymbol{\sigma}}$, where

$$\overset{(n+1)}{\boldsymbol{\sigma}} = D\overset{(n)}{\boldsymbol{\sigma}}/Dt + \overset{(n)}{\boldsymbol{\sigma}}\mathbf{V} + \left[\overset{(n)}{\boldsymbol{\sigma}}\mathbf{V}\right]^T, \qquad \overset{(0)}{\boldsymbol{\sigma}} = \boldsymbol{\sigma} \tag{4.39}$$

With this additional notation, it is now convenient to return to the constitutive functional relation (4.20). It will be recalled that, under certain restrictions, the linear *functional* (2.5) can be replaced by a *function* relation (2.16) connecting current values of stress and strain and their time derivatives. In a similar manner it has been shown (see Eringen, 1962) that (4.20) can be expressed in a more restricted form

$$\mathbf{f}\left\{\boldsymbol{\sigma}, \overset{(1)}{\boldsymbol{\sigma}}, \overset{(2)}{\boldsymbol{\sigma}}, \ldots; \mathbf{A}_1, \mathbf{A}_2, \ldots; \mathbf{B}\right\} = \mathbf{0} \tag{4.40}$$

where all arguments are evaluated at time t and $\mathbf{f}$ is a matrix *function*. The restricted class of materials described by the relation (4.40) have been called materials of rate type. In the classical linear theory of viscoelasticity $\mathbf{f}$ is a linear function, the arguments in (4.40) reduce to $\boldsymbol{\sigma}$, $\boldsymbol{\epsilon}$ and their partial time derivatives, and equation (2.16) is recovered.

A restricted class of materials which has received considerable attention in the theoretical literature is that for which $\mathbf{f}$ does not depend on the stress rates. Equation (4.40) then reduces to

$$\boldsymbol{\sigma} = \mathbf{g}\{\mathbf{A}_1, \mathbf{A}_2, \ldots; \mathbf{B}\} \tag{4.41}$$

where $\mathbf{g}$ is a matrix function. These materials have been called materials of differential type or Rivlin-Ericksen materials (Rivlin and Ericksen, 1955). When the dependence on $\mathbf{B}$ is omitted from (4.41) the equation describes the behaviour of Rivlin-Ericksen fluids. In practice, all of the mathematical models presented in this section have been found to be most appropriate for the description of viscoelastic fluids. They will not therefore play a significant part in the later chapters.

4. NONLINEAR ELASTICITY

This book is concerned with the non-linear characterization of viscoelastic solids, and the relevant extreme of the spectrum of material behaviour is that occupied by nonlinear elastic solids. For this reason it is of interest to consider the particular forms taken by the nonlinear theories of viscoelasticity when memory effects become negligible. This limiting form of the theory is also of interest because the behaviour of viscoelastic solids at the limits of the time and frequency ranges is essentially elastic. In a crosslinked material the behaviour at short times is due essentially to elastic deformation of the long-chain molecules, relative chain movement being insignificant in the short time scales of relevance. At long times and low frequencies, all possible relative chain motion has taken place, and stress depends on the elastic forces exerted by the extended molecular network.

When memory effects are neglected the various forms of functional relation discussed in Section 1 take particularly simple forms. In particular (4.34), which is obtained from (4.8), becomes

$$\boldsymbol{\sigma}(t) = \mathbf{F}(t)\,\boldsymbol{\Phi}\{\mathbf{C}(t)\}\,\mathbf{F}^T(t) \tag{4.42}$$

where $\mathbf{\Phi}$ is a matrix function. Since all functions are evaluated at time t, this dependence will not be shown explicitly in the equations of this section and (4.42) will simply be written as

$$\boldsymbol{\sigma} = \mathbf{F}\mathbf{\Phi}(\mathbf{C})\,\mathbf{F}^T. \tag{4.43}$$

The particular forms of $\mathbf{\Phi}$ appropriate to the symmetries of various crystal classes and to rotational symmetry about an axis have been given by Smith and Rivlin (1957). In the special case of an isotropic material (4.43) is expressible in the form

$$\boldsymbol{\sigma} = \mathbf{g}(\mathbf{B}) \tag{4.44}$$

which is the reduced form taken by (4.20) and (4.41), where $\mathbf{g}$ is an isotropic matrix function. Thus it is possible to express the constitutive relation for an isotropic elastic material in the form (Rivlin, 1960)

$$\boldsymbol{\sigma} = \alpha_0\,\mathbf{I} + \alpha_1\,\mathbf{B} + \alpha_2\,\mathbf{B}^2 \tag{4.45}$$

where $\mathbf{B}$ is defined in (4.21) and α_0, α_1 and α_2 are single-valued scalar functions of the strain invariants

$$I_1 = \operatorname{tr}\mathbf{B}, \qquad I_2 = \tfrac{1}{2}[(\operatorname{tr}\mathbf{B})^2 - \operatorname{tr}\mathbf{B}^2], \qquad I_3 = \det\mathbf{B} \tag{4.46}$$

where det denotes the determinant. It is readily shown that (4.46) can also be evaluated by replacing $\mathbf{B}$ by $\mathbf{C}$.

An equation of the form (4.45) is obtained from (4.26) and (4.28) when the material functions ψ_α take constant values and use is made of the Hamilton-Cayley Theorem [see equation (7.41)] together with results of the form

$$2\mathbf{R}\mathbf{E}\mathbf{R}^T = \mathbf{R}\mathbf{C}\mathbf{R}^T - \mathbf{R}\mathbf{I}\mathbf{R}^T = \mathbf{R}\mathbf{F}^T\,\mathbf{F}\mathbf{R}^T - \mathbf{I} = \mathbf{R}\mathbf{R}^T\,\mathbf{V}^T\,\mathbf{V}\mathbf{R}\mathbf{R}^T - \mathbf{I} = \mathbf{B} - \mathbf{I}. \tag{4.47}$$

For the perfectly elastic materials discussed here it is possible to express the stress in terms of a strain energy function W. In the case of isotropic materials $W = W(I_1, I_2, I_3)$ and it can be shown (see Rivlin, 1960) that

$$\alpha_0 = 2I_3^{1/2}\frac{\partial W}{\partial I_3}, \qquad \alpha_1 = 2I_3^{-1/2}\left[\frac{\partial W}{\partial I_1} + I_1\frac{\partial W}{\partial I_2}\right], \qquad \alpha_2 = -2I_3^{-1/2}\frac{\partial W}{\partial I_2} \tag{4.48}$$

In the particular case of incompressible materials the constitutive relation defines stress only to within an arbitrary hydrostatic pressure (cf. Chapter 1, Section 5). Equation (4.45) for an isotropic material is replaced by

$$\boldsymbol{\sigma} = -p\mathbf{I} + \alpha_1\,\mathbf{B} + \alpha_2\,\mathbf{B}^2 \tag{4.49}$$

and, since incompressibility requires that $I_3 \equiv 1$, the strain energy function is $W = W(I_1, I_2)$ and

$$\alpha_1 = 2\left[\frac{\partial W}{\partial I_1} + I_1\frac{\partial W}{\partial I_2}\right], \qquad \alpha_2 = -2\frac{\partial W}{\partial I_2}. \tag{4.50}$$

Experimental characterization of elastic materials reduces to the experimental determination of the strain energy function W. A number of proposed approximate forms for W, and references to experimental investigations, are contained in the paper by Rivlin (1960). A particularly simple form, which has been found to be adequate for some vulcanized rubbers, is that due to Mooney (1940):

$$W = C_1(I_1 - 3) + C_2(I_2 - 3) \tag{4.51}$$

where C_1 and C_2 are constants. When the term in I_2 is omitted, the Mooney material is described as neo-Hookean, and this is the form predicted by the simple kinetic theory of rubber-like elasticity (Treloar, 1949). When this fact is judged in the context of more general forms for W and, due to the extra complications introduced by memory effects, it is evident that the constitutive equations discussed in this book cannot be easily derived from molecular structure.

Since viscoelastic behaviour involves dissipative mechanisms the concept of a strain energy function W cannot be carried over directly from elastic to viscoelastic characterization. However, it can be replaced by a stored energy *functional* A which depends on the entire past history of deformation (see, for example, Christensen, 1971). Thus experimental characterization of viscoelastic behaviour could be considered as reducing to measurement of the scalar functional A. However, this approach has not been adopted in the literature and, in fact, experimental determination of W is achieved in practice by first determining α_0, α_1 and α_2. This does not necessarily involve additional effort since, in a representation where terms of a given order are retained, $\alpha_i(I_1, I_2, I_3)$ need to be determined to a lower order than $W(I_1, I_2, I_3)$ since the latter appears only through its derivatives. Thus, to a given order, $\partial W/\partial I_\alpha$ ($\alpha = 1,2,3$) are essentially independent functions. Similar remarks apply to the determination of viscoelastic properties either through A or directly from equations like (4.28).

5. COMPLEX MODULUS FORMULATION

Integral and differential forms of constitutive relation for nonlinear behaviour have been presented in Sections 2 and 3: formulations involving complex moduli or complex compliances remain to be considered. Because of experimental difficulties, which are discussed in Chapter 5, Section 4, accurate experimental determination of dynamic data may be too difficult in the non-linear range; no successful experimental investigation has yet been reported. For this reason the complex-function formulations have been considered only in one-dimensional form (Nakada, 1960; Lockett and Gurtin, 1964): extension to three dimensions is straightforward but tedious.

Consider a one-dimensional integral constitutive relation connecting a single stress component σ with a single strain component E

$$\sigma(t) = \sum_{n=1}^{N} \sigma_n(t) \tag{4.52}$$

where

$$\sigma_n(t) = \int_0^\infty \dots \int_0^\infty \Phi_n(s_1, s_2, \dots, s_n)\, \dot{E}(t - s_1) \dots \dot{E}(t - s_n)\, ds_{(n)}$$

$$ds_{(n)} = ds_1\, ds_2 \dots ds_n. \tag{4.53}$$

This constitutive relation is a special one-dimensional form of (4.26) and (4.28); the conditions under which it is applicable are discussed in Chapter 5. A dual theory could also be considered in which the roles of σ and E are interchanged and, in that case, a complex compliance formulation would result.

The one-sided complex Fourier transforms of the stress-relaxation functions Φ_n are defined by

$$\bar{\Phi}_n(\omega_1, \omega_2, \dots, \omega_n) = \int_0^\infty \dots \int_0^\infty \Phi_n(s_1, s_2, \dots, s_n) \exp[-i(\omega_1 s_1 + \omega_2 s_2 + \dots + \omega_n s_n]\, ds_{(n)} \tag{4.54}$$

and, by analogy with (2.21), it is convenient to define a set of complex moduli E_n by

$$E_n(\omega_1, \omega_2, \dots, \omega_n) = i^n \omega_1 \omega_2 \dots \omega_n \bar{\Phi}_n(\omega_1, \omega_2, \dots, \omega_n). \tag{4.55}$$

The complex Fourier transforms of stress and strain are defined by

$$[\bar{\sigma}(\omega), \bar{E}(\omega)] = \int_{-\infty}^{\infty} [\sigma(t), E(t)] \exp(-i\omega t)\, dt \tag{4.56}$$

for which the inversion formulae are

$$[\sigma(t), E(t)] = \frac{1}{2\pi} \int_{-\infty}^{\infty} [\bar{\sigma}(\omega), \bar{E}(\omega)] \exp(i\omega t)\, d\omega. \tag{4.57}$$

Substitution of (4.57) into (4.53), interchange of the orders of integration, and use of (4.54) and (4.55), now results in the expression

$$\sigma_n(t) = (2\pi)^{-n} \int_{-\infty}^{\infty} \cdots \int_{-\infty}^{\infty} E_n(\omega_1, \omega_2, \ldots, \omega_n)\, \bar{E}(\omega_1)\, \bar{E}(\omega_2) \ldots \bar{E}(\omega_n)$$

$$\times \exp\left[i(\omega_1 + \omega_2 + \cdots + \omega_n)\, t\right] d\omega_{(n)}. \tag{4.58}$$

When $n = 1$ equations (4.56) and (4.58) show that

$$\bar{\sigma}_1(\omega) = E_1(\omega)\, \bar{E}(\omega) \tag{4.59}$$

which, apart from some differences in notation, is equivalent to the form (2.20) used in the classical linear theory. For $n > 1$ equations (4.56) and (4.58) give

$$\bar{\sigma}_n(\omega) = (2\pi)^{-n+1} \int_{-\infty}^{\infty} \cdots \int_{-\infty}^{\infty} E_n(\omega_1, \omega_2, \ldots, \omega_{n-1}, p_n)$$

$$\times \bar{E}(\omega_1)\, \bar{E}(\omega_2) \ldots \bar{E}(\omega_{n-1})\, \bar{E}(p_n)\, d\omega_{(n-1)} \tag{4.60}$$

where

$$p_n = \omega - (\omega_1 + \omega_2 + \cdots + \omega_{n-1}) \tag{4.61}$$

and use has been made of the formal result

$$\int_{-\infty}^{\infty} \exp\left[i(\omega_1 + \cdots + \omega_n - \omega)\, t\right] dt = 2\pi\delta(\omega_1 + \cdots + \omega_n - \omega) \tag{4.62}$$

where δ is the Dirac delta. Equations (4.59) and (4.60), together with the transform of (4.52):

$$\bar{\sigma}(\omega) = \sum_{n=1}^{N} \bar{\sigma}_n(\omega) \tag{4.63}$$

give the complex modulus form of the one-dimensional constitutive relation. Unlike the linear form (2.20), or the first-order form (4.59), the nonlinear relation is an integral relation rather than a purely algebraic expression connecting $\bar{\sigma}$ and $\bar{E}$. Thus a correspondence principle relating elastic and viscoelastic theories does not exist in the nonlinear theory.

Chapter 5 One-Dimensional Green–Rivlin Behaviour

It was shown in Chapter 3 that linear viscoelastic properties can be conveniently measured by considering certain simple types of deformation for which the three-dimensional constitutive relation reduces essentially to a one-dimensional form. By considering a number of such simple cases the full three-dimensional behaviour can be determined. Two experiments are required for an isotropic material, and simple shear and extension tests are the most convenient: in stress-relaxation tests material functions are determined which correspond to the shear modulus and Young's modulus of an elastic material. If the extension test is treated as a three-dimensional deformation, and the lateral contraction is measured in addition to the longitudinal extension, then this test alone is adequate in principle to determine the material properties. However, experimental difficulties of accurate measurement of the lateral contraction may make it desirable to perform the extra shear test, at least as a check on the accuracy of the data.

Similar remarks apply to the experimental determination of material functions for characterizing nonlinear behaviour. The complete set of experiments which is required for determining the three-dimensional properties of an isotropic material is given in Chapter 6. In some cases these experiments are essentially one-dimensional, but in other cases they require deformations and loads to be measured or applied in more than one direction. This requirement, which is not essential for linear behaviour, is due to the nonlinear nature of interactions between different load components. The present chapter is concerned with purely one-dimensional behaviour; this may be of interest in its own right in certain practical applications, such as stretching of fibres or may be considered as a special case of the three-dimensional theory given in Chapter 6.

The one-dimensional form of the Green-Rivlin theory is presented in Section 1, and it is shown how this relation may arise in special types of deformation. A set of creep or stress-relaxation tests is described in Section 2, sufficient to determine all of the material functions appearing in this theory. Experimental requirements of simplified theories, which depend on additional physical or mathematical assumptions, are treated in later chapters. A number of experimental investigations have attempted to use the Green-Rivlin theory to model the creep behaviour of real materials and these are described in Section 3.

The final section treats the problem of material property determination by oscillatory tests.

1. ONE-DIMENSIONAL CONSTITUTIVE RELATION

A. Integral Polynomial Form

The constitutive relation is considered in the form

$$Q(t) = \int_{-\infty}^{t} \Phi_1(t-\tau_1)\dot{P}(\tau_1)\,d\tau_{(1)} + \iint_{-\infty}^{t} \Phi_2(t-\tau_1, t-\tau_2)\dot{P}(\tau_1)\dot{P}(\tau_2)\,d\tau_{(2)}$$

$$+ \iiint_{-\infty}^{t} \Phi_3(t-\tau_1, t-\tau_2, t-\tau_3)\dot{P}(\tau_1)\dot{P}(\tau_2)\dot{P}(\tau_3)\,d\tau_{(3)} + \cdots \qquad (5.1)$$

where $d\tau_{(N)} = d\tau_1\,d\tau_2\ldots d\tau_N$ and the Φ_n are functions which can be assumed, without loss in generality, to be symmetric with respect to their indicated arguments. $P(t)$ and $Q(t)$ denote stress and strain components in a one-dimensional relation; possible precise definitions of these quantities and justification of (5.1) is considered below. Since it is convenient in the interpretation of an experimental investigation to apply special forms of function $P(t)$ and measure the resulting function $Q(t)$, these functions will be referred to as input and output functions, respectively. If $P(t)$ denotes a strain measure, the material functions Φ_n are stress-relaxation functions of order n for the mode of deformation relevant to the definitions of P and Q. If $P(t)$ denotes stress then the Φ_n are creep functions.

In the general theory the integral polynomial (5.1) is of infinite order. However, it must obviously be terminated at some point when used in practice. Here it is assumed that the terms written explicitly in (5.1) are adequate; it will be seen later that the experimental task associated with this relation is considerable. Since the signs of shear components depend on the choice of axes used to define the deformation, shear behaviour is essentially an odd-order phenomenon; this can be readily verified from the general theory (4.28). It follows that a third-order theory is the simplest form which is potentially capable of modelling nonlinear behaviour in all modes of deformation. For some materials and ranges of loading, a third-order relation may be inadequate and further terms must then be included in the constitutive relation. Extension of the theoretical basis for the experimental programme is straightforward, but the experimental programme would soon become prohibitive. When this occurs the general theory must be abandoned and a special relation sought similar to those discussed in Chapter 9.

B. Deformations for which $\mathbf{R} = \mathbf{I}$

Consider extension of a rod or fibre of material, in which the longitudinal applied *stress* is denoted by $P(t)$, all other stress components being zero. In this case the matrix $\mathbf{M}$ appearing in (4.28) is given by

$$\mathbf{M}(\tau) = \begin{Vmatrix} \dot{P}(\tau) & 0 & 0 \\ 0 & 0 & 0 \\ 0 & 0 & 0 \end{Vmatrix} \tag{5.2}$$

since $\mathbf{R} = \mathbf{I}$, and where the x_1-direction has been chosen as the axis of the rod or fibre. When (5.2) is substituted into (4.28) there results a set of equations for the non-zero components of $\mathbf{Q}(t)$ (that is, Q_{11} and $Q_{22} = Q_{33}$) each of which is of the form (5.1); the Φ_n are simple combinations of the material functions ψ_α. Thus, for this situation, (5.1) may represent an expression for the longitudinal extension or lateral contraction (the Φ_n are different functions in the two cases); the former is likely to be of greater practical value.

Provided that inertia effects can be neglected, the homogeneous deformation field produced by the applied load is of the form

$$x_1 = X_1 + f(t)\, X_1, \qquad x_2 = X_2 + g(t)\, X_2, \qquad x_3 = X_3 + g(t)\, X_3 \tag{5.3}$$

where the origin of coordinates is at the stationary point, and $f(t)$ and $g(t)$ depend on $P(t)$ through (5.1). The longitudinal components of the classical and finite strain measures are

$$\epsilon_{11} = f(t), \qquad E_{11} = f(t) + \tfrac{1}{2}[f(t)]^2 \tag{5.4}$$

so that

$$E_{11} = \epsilon_{11} + \tfrac{1}{2}\epsilon_{11}^2. \tag{5.5}$$

It follows that not only E_{11} but also the classical strain ϵ_{11} can be represented by an expression of the form (5.1). However, the creep functions will be different in the two cases, and the two expressions of third order in $P(t)$ represent different approximations of the general integral polynomial relation.

The example discussed above refers to a test in which the input function $P(t)$ is a stress, and experiments are therefore of creep type, the functions Φ_n denoting creep functions. Tests of stress-relaxation type, in which the input function is a strain, can be treated in a similar manner. However, it is difficult experimentally to restrict the lateral contraction to zero and more convenient to impose some deformation in the longitudinal direction and to allow the lateral deformation to take on a form consistent with zero lateral loading. Thus, it is supposed that the imposed longitudinal extension and the transverse contraction due to free lateral loading are such that

$$E_{11}(\tau) = P(\tau), \qquad E_{22}(\tau) = E_{33}(\tau) = A(\tau) \tag{5.6}$$

where $A(\tau)$ is unknown and inertia effects will be neglected. When (5.6) is substituted into (4.28) there results an equation of the form

$$Q_{11}(t) = \int_{-\infty}^{t} [\phi_1 \dot{P}(\tau) + \phi_2 \dot{A}(\tau)]\, d\tau + \iint_{-\infty}^{t} [\phi_3 \dot{P}(\tau_1)\dot{P}(\tau_2) + \phi_4 \dot{P}(\tau_1)\dot{A}(\tau_2)$$
$$+ \phi_5 \dot{A}(\tau_1)\dot{A}(\tau_2)]\, d\tau_{(2)} + \iiint_{-\infty}^{t} \{\cdots\}\, d\tau_{(3)} \tag{5.7}$$

where the ϕ_α are stress-relaxation functions: ϕ_1 and ϕ_2 are functions of $t-\tau$; ϕ_3, ϕ_4 and ϕ_5 are functions of $t-\tau_1$ and $t-\tau_2$; etc. Since the deformation is such that $\mathbf{R} = \mathbf{I}$ it follows that $Q_{11} = \sigma_{11}$ and (5.7) is an explicit expression for the longitudinal stress; a similar expression is readily obtained for the lateral stresses $\sigma_{22} = \sigma_{33} = 0$.

A relation expressing $Q_{11}(t)$ as a function of the input $P(\tau)$ alone can, in principle, be obtained by eliminating A between the expressions for Q_{11} and Q_{22}. Although this may result in an expression of the form (5.1), it is not an easy matter to relate the material functions in that equation with the functions ψ_α appearing in the full three-dimensional equation. One trivial exception is the case of a linear material, where elimination of A is a simple matter. A second and more useful exception is the case of an incompressible material for which the function $A(t)$ can be calculated from $P(t)$ using the condition of zero volume change. Thus both Q_{11} and $Q_{22} = 0$ can be expressed as explicit functions of $P(\tau)$ and it is only necessary to eliminate between these expressions the arbitrary hydrostatic pressure which is a feature of constitutive relations for incompressible materials (see Chapter 6 for further details).

Although extension of a rod or fibre is the simplest experimental configuration leading to an equation of the form (5.1), there are other forms of deformation which are characterized by an equation of this type. In a homogeneous biaxial test a sheet of material is subjected to equal uniform loading in two perpendicular directions in its plane: $\sigma_{11} = \sigma_{22} = \sigma$, and the other stress components are identically zero. In this case $\mathbf{R} = \mathbf{I}$ and E_{33} and $E_{11} = E_{22} = E$ are the only non-zero strain components. If creep tests are considered $\mathbf{M} = \text{diag}\{\dot{\sigma}, \dot{\sigma}, 0\}$ and (4.28) gives an expression of the form (5.1) for E: it also gives a similar expression for E_{33} but this will have little experimental value. If stress-relaxation tests are considered (4.28) gives an expression for σ in terms of E and E_{33}, and the latter must be eliminated using the condition that $\sigma_{33} = 0$. As for the case of simple extension, this elimination is more readily performed for an incompressible material.

Another situation leading to (5.1), and also involving deformations such that $\mathbf{R} = \mathbf{I}$, is bulk behaviour of materials. Here $\sigma_{11} = \sigma_{22} = \sigma_{33} = \sigma$, say, and

$E_{11} = E_{22} = E_{33} = E$ say, and (4.28) readily yields an expression of the form (5.1) connecting σ and E. Although the mathematical representation of this type of experiment has attractions of symmetry and simplicity, actual experiments on bulk behaviour are difficult to perform. The author knows of no experimental data on nonlinear viscoelastic properties in bulk behaviour. For an incompressible material this experiment is meaningless, since E cannot then be made non-zero.

C. Deformations for which $\mathbf{R} \neq \mathbf{I}$

Apart from the experiments discussed above, some testing procedures involve local rotation of the material in which case $\mathbf{R} \neq \mathbf{I}$. A convected or rotated stress matrix is involved in the constitutive relation and, if this is to be used directly, it is necessary either (i) to apply the load in a creep test so as to maintain the convected or rotated stress matrix at a constant value, or (ii) to measure or calculate the stresses in a stress-relaxation test in the convected or rotated system. The former involves very difficult experimental problems, as does direct measurement in the latter case. Although calculation in the latter case seems to be plausible, it should be realized that this requires knowledge of all (or certainly more than one) component of the stress matrix measured in the fixed coordinate system. For these reasons direct use of a constitutive relation involving the convected or rotated stress matrix have not been used so far to analyse experimental data, however, recourse has been made to the simplifications offered by certain approximations.

Lifshitz (1964) and Lifshitz and Kolsky (1967) considered torsion of a thin-walled hollow circular cylinder whose axis may be taken parallel to the x_1-direction. At a typical material point the x_2 and x_3-axes are taken in the circumferential and radial directions respectively. The rotation matrix $\mathbf{R}$ now has the approximate form

$$\mathbf{R} = \begin{Vmatrix} 1 & -\dfrac{\gamma}{2} & 0 \\ \dfrac{\gamma}{2} & 1 & 0 \\ 0 & 0 & 1 \end{Vmatrix} \tag{5.8}$$

where γ is the angle through which a generator of the cylinder has been rotated, and this approximate form holds for sufficiently small γ. If the only non-zero stress component $\sigma_{12} = \sigma_{21}$ is denoted by $\bar{\tau}$ then $\mathbf{P}$ can be expressed as

$$\mathbf{P} = \mathbf{R}^T \boldsymbol{\sigma} \mathbf{R} = \begin{Vmatrix} \bar{\tau}\gamma & \bar{\tau} & 0 \\ \bar{\tau} & -\bar{\tau}\gamma & 0 \\ 0 & 0 & 0 \end{Vmatrix} \tag{5.9}$$

where further use has been made of the assumption $\gamma \ll 1$.

The following derivation deviates slightly from that used by Lifshitz and Kolsky. It is convenient to start from a relation of the form (4.28) in which $\mathbf{M}_\alpha$ is replaced by $\mathbf{P}_\alpha$. It has been noted that this is possible using integration by parts, the creep functions being replaced by suitable derivatives which can equally well be used as material functions. Using the notation $\bar{\tau}_\alpha = \bar{\tau}(\tau_\alpha)$, $\gamma_\alpha = \gamma(\tau_\alpha)$ it is found that

$$(\mathbf{P}_1)_{12} = \bar{\tau}_1, \qquad (\mathbf{P}_1 \mathbf{P}_2)_{12} = (\gamma_1 - \gamma_2)\,\bar{\tau}_1\,\bar{\tau}_2$$

$$(\mathbf{P}_1 \mathbf{P}_2 \mathbf{P}_3)_{12} = \bar{\tau}_1\,\bar{\tau}_2\,\bar{\tau}_3, \qquad T_{12}(\mathbf{P}_3)_{12} = 2\bar{\tau}_1\,\bar{\tau}_2\,\bar{\tau}_3 \tag{5.10}$$

where $(\quad)_{12}$ denotes the 12-component of a matrix, and $\gamma_1\gamma_2$, etc. have been neglected in comparison with unity. When expressions (5.10) are substituted into the expression for $E_{12} = \gamma/2$ obtained from the three-dimensional relation there results the equation

$$\tfrac{1}{2}\gamma(t) = \int_{-\infty}^{t} \bar{\psi}_2\,\bar{\tau}(\tau_1)\,d\tau_1 + \iiint_{-\infty}^{t} (2\bar{\psi}_{10} + \bar{\psi}_{12})\,\bar{\tau}(\tau_1)\,\bar{\tau}(\tau_2)\,\bar{\tau}(\tau_3)\,d\tau_{(3)} \tag{5.11}$$

where the second-order contribution vanishes due to symmetry of ψ_6. A bar has been placed over the material functions to indicate that they are not the functions appearing in (4.28), but are those appearing in the relation in which $\mathbf{M}$ is replaced by $\mathbf{P}$. By a repeated use of integration by parts (5.11) can be converted to a form involving $\dot{\bar{\tau}}(\tau_\alpha)$ in the integrands, and this equation is a special case of (5.1).

It is interesting to note that (5.11) is identical to the form that is obtained by setting $\mathbf{R} = \mathbf{I}$ in place of (5.8); this is due to neglect of $\gamma_1\gamma_2$, etc. in comparison with unity and applies only to the relation for the shear component E_{12}. The expressions for E_{11}, E_{22} and E_{33} contain terms linear in γ in their integrands and are not precisely of the form (5.1).

A number of possible situations which can be represented by the one-dimensional constitutive relation (5.1) have been described and attention will now be given to the problem of determining the functions Φ_α appearing in (5.1) from experimental data.

2. CREEP AND STRESS-RELAXATION EXPERIMENTS

Depending on whether the input function $P(\tau)$ is a stress or a strain, the experiments described below are of creep or stress-relaxation type. No distinction is made here between the two alternatives; the theoretical motivation is the same in both cases. In a simple creep or stress-relaxation test such as those considered in Section 1, Chapter 3, the input function is

$$P(t) = pH(t) \tag{5.12}$$

where p is a constant and H denotes the Heaviside step function. When this expression is substituted into (5.1) it follows that

$$Q(t) = p\Phi_1(t) + p^2\Phi_2(t, t) + p^3\Phi_3(t, t, t) \tag{5.13}$$

and the measured response $Q(t)$ from one experiment alone is not sufficient to determine any of the material functions. However, if the simple test (5.12) is performed for three different values of p, then (5.13) and the measured outputs provide three equations for determining

$$\Phi_1(t), \qquad \Phi_2(t, t), \qquad \Phi_3(t, t, t). \tag{5.14}$$

Thus Φ_1 is known over the time scale of the experiments, but Φ_2 and Φ_3 are known only when their arguments are all equal. To determine further values of these functions it is necessary to consider an input

$$P(t) = p_1 H(t) + p_2 H(t - k) \tag{5.15}$$

where p_1, p_2 and k are constants. This represents an input with sudden increases at $t = 0$ and $t = k$, and it produces a response

$$Q(t) = 2p_1 p_2 \Phi_2(t, t - k) \\ + 3p_1 p_2[p_1 \Phi_3(t, t, t - k) + p_2 \Phi_3(t, t - k, t - k)] \quad + \cdots \tag{5.16}$$

where the terms not written explicitly in (5.16) depend only on the known functions (5.14). In (5.16) use has been made of the assumption that the Φ_α are symmetric with respect to their arguments. If three experiments are performed, corresponding to three independent combinations of the parameters p_1 and p_2, equation (5.16) allows the measured responses to be used to determine

$$\Phi_2(t, t - k), \qquad \Phi_3(t, t, t - k), \qquad \Phi_3(t, t - k, t - k). \tag{5.17}$$

Values of the function Φ_2 have now been determined along those lines in the argument plane indicated by full lines in Fig. 1. (Note that Φ_2 has been chosen to be symmetric in its arguments, and all of the material functions vanish when any argument is negative; this latter statement expresses the fact that responses to loads do not occur before the load is applied.) If the set of three experiments (5.11) is now repeated for other values of k, Φ_2 can be determined along a sufficient number of lines in the argument plane (shown as broken lines in Fig. 1) for adequate determination of Φ_2 at any point by interpolation. In common with the difficulties noted in the linear case it is not possible to extrapolate the data precisely for zero and infinite times. The number N of values of k which is required, and the precise values of k, are best determined on an *ad hoc* basis depending on the measured experimental data; it may be expected that N will be of the order of ten for reasonable characterization.

Similar remarks apply to the values of Φ_3 obtained from (5.17); in this case the function is determined along lines in its three-dimensional argument space. The two values of Φ_3 indicated in (5.17) lead to determination of Φ_3 when two of its arguments are equal, the coincident arguments being greater than, or less than, the third argument respectively. Thus as a result of $3N$ experiments of the type (5.15) the function Φ_2 is determined completely and the function Φ_3 is

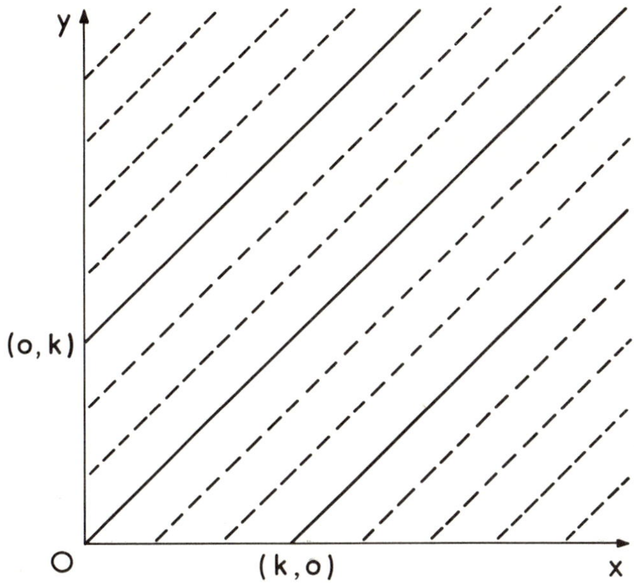

Fig. 1. Argument plane of the function $\Phi_2(x,y)$. Experiments governed by equations (5.12) and (5.15) determine Φ_2 along unbroken lines. Values on broken lines are determined from other values of k.

determined except when its arguments are all unequal. To obtain these unknown values consider an input of the form

$$P(t) = p_1 H(t) + p_2 H(t-k) + p_3 H(t-l) \tag{5.18}$$

where p_1, p_2, p_3, k and l are constants; (5.14) corresponds to a triple-step input. Equation (5.1) now gives the output

$$Q(t) = 6p_1 p_2 p_3 \Phi_3(t, t-k, t-l) + \cdots \tag{5.19}$$

where the terms not written explicitly are known functions determined in (5.14) and (5.17). Thus the single experiment (5.18) leads to determination of

$$\Phi_3(t, t-k, t-l) \tag{5.20}$$

and repetition of the experiment for other values of k and l determines Φ_3 completely. Due to symmetry of Φ_3 it is not necessary to repeat the experiment for N values of both k and l. It is sufficient to consider $0 < k < l$. The total number of experiments of type (5.18) is therefore $\frac{1}{2}N(N-1)$. The complete set of experiments required for experimental determination of the Φ_α is summarized in Table 1.

It should be noted that the programme described above represents the *minimum* requirements for determining the functions Φ_α. Many experiments are required but these are analysed in groups of three at most. Even so, solution of three simultaneous equations of the form (5.13) or (5.16) would require a high degree of experimental accuracy in order to determine the Φ_α with any confidence. Thus, in practice, it would be preferable to repeat each experiment a number of times and determine average values for measured quantities. In general, this leads to a large experimental programme as the numbers discussed below will indicate.

Table 1. Experimental programme for determination of Φ_α (after Lockett, 1965)

Input	Number of independent choices of p_i	Number of choices of k and l	Total number of experiments of type
1. $P(t) = pH(t)$	3	1	3
2. $P(t) = p_1 H(t) + p_2 H(t-k)$	3	N	$3N$
3. $P(t) = p_1 H(t) + p_2 H(t-k)$ $+ P_3 H(t-l)$	1	$\frac{1}{2}N(N-1)$	$\frac{1}{2}N(N-1)$
		Total	$\frac{1}{2}(N+3)(N+2)$

Possibly the smallest value of N for which a crude characterization could be obtained is $N = 5$, and in this case the total number of experiments is 28. When N takes the more satisfactory value of 10, this number becomes 78. If each of these experiments is to be repeated several times to improve data accuracy through averaging, the number of experiments becomes prohibitive. It appears that, in its general form, this method of material characterization could be useful as a research tool to investigate a few selected materials, but is too complex to be used as a standard measurement procedure; possible simpler forms of characterization are discussed in later chapters. However, it is not necessary to complete the entire experimental programme in order to obtain some useful information about material behaviour. It is noted that the $3(N+1)$ experiments of types (5.12) and (5.15) allow determination of all function values except those for Φ_3 when all of its arguments are unequal. Thus it is seen that, in the case $N = 5$, only 18 experiments are needed to obtain a significant amount of information. The main value of a general discussion such as that given above is the fact that it permits individual or incomplete sets of data to be interpreted in the context of the information they give towards a complete characterization.

The experiments discussed in the present section are multiple-step creep or stress-relaxation tests. By comparison with the linear theory it is to be expected that other experimental techniques, such as oscillatory measurements, could be used to determine nonlinear mechanical properties. This possibility will be discussed in Section 4, but this is preceded by a review of experimental investigations which have used the methods discussed in the present section.

3. EXPERIMENTAL INVESTIGATIONS

No experimental investigation has included all of the experiments discussed in the previous section, even for a single material. However, there are now a significant number of sets of data which go some way towards determining the material functions in a relation like (5.1) for certain materials; it will be convenient to discuss the data and conclusions obtained from a number of different types of test.

A. Single-Step Creep Test

In this case (5.13) applies in which Q is a strain component and p denotes a stress magnitude. For extension of a rod or fibre

$$E(t) = \sigma_0 \Phi_1(t) + \sigma_0^2 \Phi_2(t,t) + \sigma_0^3 \Phi_3(t,t,t) \tag{5.21}$$

where E is the longitudinal strain and σ_0 is the magnitude of the applied longitudinal stress. It has been noted earlier that (5.21) could be replaced by a similar expression in which E is replaced by ϵ; here the form (5.21) is considered in order to discuss the experimental data of Lifshitz (1964) and Lifshitz and Kolsky (1967). These authors investigated the behaviour of polyethylene at stress levels in the range 200 to 800 lb/in^2; in the time scale of the experiments (100 min) this corresponded to a maximum strain E of 10 %. Presentation of data is simplified by defining dimensionless quantities

$$\sigma_* = \sigma_0/\bar{\sigma}, \qquad \Phi_\alpha^* = \bar{\sigma}^\alpha \Phi_\alpha \tag{5.22}$$

where $\bar{\sigma}$ is a typical stress level. Equation (5.21) then becomes

$$E(t) = \sigma_* \Phi_1^*(t) + \sigma_*^2 \Phi_2^*(t,t) + \sigma_*^3 \Phi_3^*(t,t,t) \tag{5.23}$$

and Lifshitz and Kolsky's tensile results for polyethylene are shown by the unbroken curves in Fig. 2, where $\bar{\sigma} = 10^3$ lb/in^2. It is interesting to note that all three functions increase in magnitude with increasing time, but the second-order function $\Phi_2(t,t)$ is negative over the whole range. At the lower end of the experimental stress range, $\sigma_* \simeq 0.2$ and it is readily seen from (5.23) and Fig. 2. that the first-order term produces the dominant contribution; in fact the small contributions from Φ_2^* and Φ_3^* are opposite in sign and almost cancel. At the highest stress level investigated $\sigma_* \simeq 0.8$ and the contributions from all three terms are comparable; thus the behaviour is significantly nonlinear at these

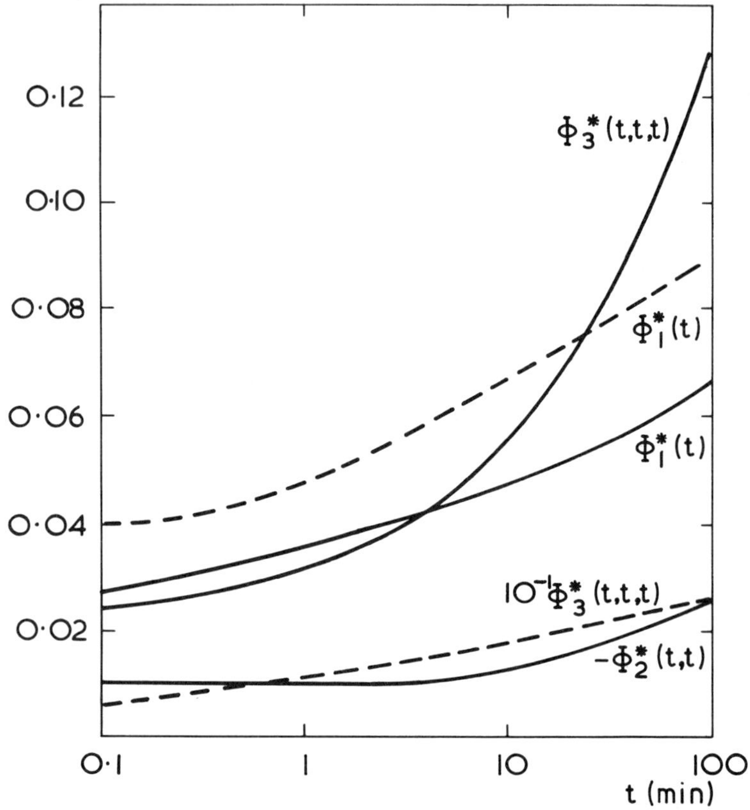

Fig. 2. Creep functions for polyethylene in tension (unbroken curves) and torsion (broken curves), after Lifshitz and Kolsky (1967).

stress levels. Predictions obtained from (5.23) and the data of Fig. 2 agreed well with experimental data for all stresses in the range considered which confirms the adequacy of a third-order relation for describing the data.

In an investigation, discussed in greater detail in sub-sections B and C, Ward and Onat (1963) found that their extensional data for polypropylene fibres could be modelled satisfactorily by a relation in which $\Phi_2 \equiv 0$. It is evident from Fig. 2 that the second-order term gives a significant contribution for polyethylene at the higher stress levels, and the data of Lifshitz and Kolsky (1967) cannot be represented by a model containing only first- and third-order terms. Thus the result of Ward and Onat (1963), which may eventually be given a physical explanation for polypropylene, is not a universal result.

In Section 1 it was shown that torsion of a thin-walled circular cylinder can be described by a relation of the form (5.1), where $\Phi_2 \equiv 0$ provided that $\gamma \ll 1$. For this case it is possible to replace E in (5.23) by E_{12}, set $\Phi_2^* = 0$, and interpret σ_* as a ratio of shear stresses. Corresponding values for Φ_1^* and Φ_3^* obtained from Lifshitz and Kolsky's (1967) data for polyethylene are shown as broken lines in Fig. 2 where $\bar{\sigma} = 10^3$ lb/in^2 and experimental stress levels were in the range 100 to 550 lb/in^2. The maximum shear strain E_{12} attained after 100 min at the highest stress level corresponded to a value $\gamma = 0.17$ which is at the limit of validity of the assumption $\gamma \ll 1$. It is readily seen that, at the lowest stress levels investigated, the behaviour is essentially linear, whereas at the higher stress levels the contribution from the third-order term is significant.

B. Two-Step Creep Tests

The most common forms of double-step test which have been reported in the literature are experiments of creep type in which the second step corresponds to removal or doubling of the first load. These loading histories are shown in Fig. 3 together with the notation which will be used for the strain response during the creep, recovery and additional-load phases. With this notation, and using equation (5.1) it is readily shown that

$$E_c(\sigma_0, t) = \sigma_0 \Phi_1(t) + \sigma_0^2 \Phi_2(t, t) + \sigma_0^3 \Phi_3(t, t, t) \tag{5.24}$$

$$\begin{aligned} E_r(\sigma_0, k; t) = E_c(\sigma_0, t-k) &- 2\sigma_0^2[\Phi_2(t-k, t-k) - \Phi_2(t, t-k)] \\ &+ 3\sigma_0^3[\Phi_3(t, t, t-k) - \Phi_3(t, t-k, t-k)] \end{aligned} \tag{5.25}$$

$$\begin{aligned} E_d(\sigma_0, k; t) = E_c(\sigma_0, t-k) &+ 2\sigma_0^2 \Phi_2(t, t-k) \\ &+ 3\sigma_0^3[\Phi_3(t, t, t-k) + \Phi_3(t, t-k, t-k)]. \end{aligned} \tag{5.26}$$

It should be noted that E_r and E_d denote the reduction or addition to the strain that would have existed if the first loading step had remained undisturbed. Because of the choice of origin of the time scale, E_r and E_d are undefined for $t < k$. It will be seen that, in the linear theory, E_r and E_d are both given by the creep curve E_c translated through a time interval k.

In their experimental investigation of the extensional behaviour of oriented polypropylene fibres, Ward and Onat (1963) noted that isochronous plots of

$E_c(\sigma_0, t)/\sigma_0$ against σ_0 could all be modelled adequately by an expression of the form (5.24) in which $\Phi_2 \equiv 0$. In addition they investigated the implications of this observation on the agreement between predictions (5.25) and (5.26) and their own experimental data, and found satisfactory agreement. In this case (5.25) and (5.26) yield

$$E_d(\sigma_0, k; t) - E_r(\sigma_0, k; t) = 6\sigma_0^3 \Phi_3(t, t-k, t-k)$$

$$E_d(\sigma_0, k; t) + E_r(\sigma_0, k; t) - 2E_c(\sigma_0, t-k) = 6\sigma_0^3 \Phi_3(t, t, t-k) \qquad (5.27)$$

so that the quantities on the left of (5.27) should both vary like σ_0^3 in isochronous plots; the two coefficients determine $\Phi_3(t, t, t-k)$ and $\Phi_3(t, t-k, t-k)$.

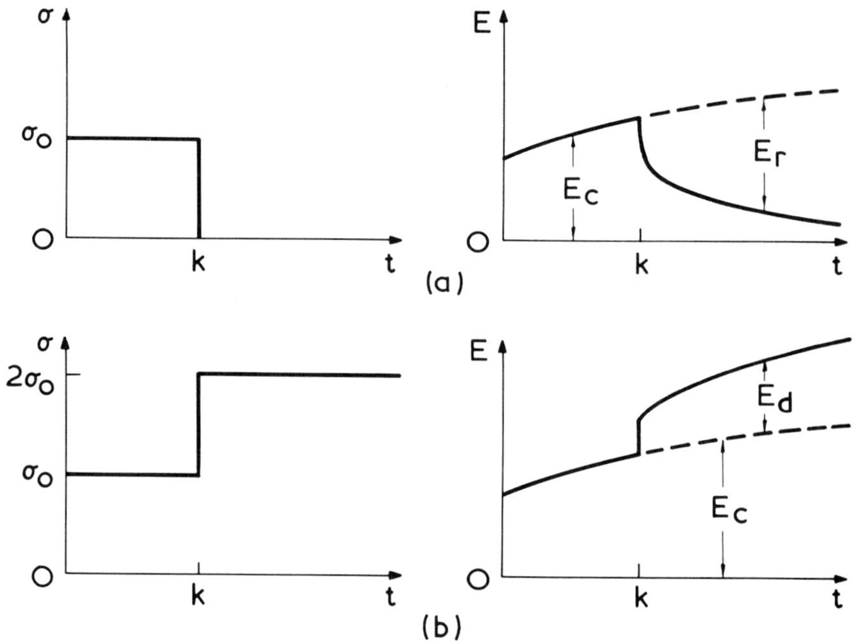

Fig. 3. Notation for strain response in (a) creep and recovery, (b) creep and additional loading.

Ward and Onat's (1963) data indicate that, when all curves are translated to a common time origin,

$$E_d > E_r > E_c \qquad (5.28)$$

so that (5.27) implies that $\Phi_3(t, t, t-k) > 0$, $\Phi_3(t, t-k, t-k) > 0$. Their experimental data also indicate that $E_d(\sigma_0, k_1; t) > E_d(\sigma_0, k_2; t)$ when $k_1 < k_2$; this implies that a material becomes stiffer with respect to additional loads when the first load is applied for a longer time. This finding is in agreement with that of Turner (1966) who investigated a propylene homopolymer; some of his results reinforce and extend those of Ward and Onat (1963).

Further experimental investigation of a model with $\Phi_2 \equiv 0$ for polypropylene has been undertaken by Ward and Wolfe (1966). Their data confirm the validity of the model for describing creep, creep-and-recovery, and superposition of identical loads, and gives further confirmation of the inequality (5.28). In addition they considered loading histories

$$\sigma(\tau) = \sigma_1 H(\tau) + \sigma_2 H(\tau - k) \tag{5.29}$$

where $\sigma_1 > 0$, $\sigma_2 > 0$ and $\sigma_1 \neq \sigma_2$; experimental data were obtained for the pairs $(\sigma_1, \sigma_2) = (\sigma_0, 2\sigma_0), (2\sigma_0, \sigma_0), (\sigma_0, 3\sigma_0), (3\sigma_0, \sigma_0)$. For the first of these pairs it is readily shown that the strain response $E(\sigma_0, 2\sigma_0, k; t)$ is given by

$$\begin{aligned} E(\sigma_0, 2\sigma_0, k; t) = E_c(\sigma_0, t) + E_c(2\sigma_0, t-k) + 3E_d(\sigma_0, k; t) \\ - 2E_c(\sigma_0, t-k) - E_r(\sigma_0, k; t) \end{aligned} \tag{5.30}$$

where use has been made of (5.27); similar expressions can be derived for the other (σ_1, σ_2) pairs. In the linear theory (5.27) shows that $E_c(\sigma_0, t-k) = E_r(\sigma_0, k; t) = E_d(\sigma_0, k; t)$ and, in that case, the last three terms in (5.30) cancel.

Ward and Wolfe (1966) compared their experimental data for the various (σ_1, σ_2) pairs with the values predicted by (5.30) and similar expressions, and with the corresponding reduced equations in which the last three terms of (5.30) are omitted. Since these terms cancel in the linear theory, this comparison indicates the significance of the nonlinear terms. It was found that their experimental data at $t = k + 100$ sec, which involved strains up to 1.8% and $k = 10^4$ sec, agreed with the strains predicted by (5.30) to within 4% of the measured values whereas the reduced form agreed only to within 8%. These results give a misleading impression of the degree of nonlinearity exhibited by the material. In fact the curves E_r and E_d differ (at $t = k + 100$ sec) from E_c by as much as 13% and 37% respectively, but the effect of these discrepancies is diminished in forming the combination required by (5.30). Thus, although the behaviour is markedly nonlinear, the reduced theory provides a reasonable approximation, and this accounts for the relative ease of modelling behaviour in loading programmes involving increasing loads. It will be seen in Chapter 9 that behaviour for non-monotonic histories is more troublesome.

C. Order of Integral Polynomial

Hadley and Ward (1965) investigated the creep and recovery behaviour of several polypropylene fibres having a range of molecular structures. Two molecular weights were produced, and within each class various degrees of molecular orientation were obtained. Although the investigation is by no means conclusive, it is possible to draw some tentative conclusions regarding the relationship between molecular structure and mechanical behaviour.

For each of the fibres investigated creep and recovery experiments were performed with $k = 120$ min and applied loads such that the maximum strain was in the range 3–5%. It will be recalled that, for the monofilaments considered by Ward and Onat (1963) and Ward and Wolfe (1966), the isochrones of E_c/σ_0 and E_r/σ_0 versus σ_0 could be adequately represented by parabolas. This fact implies that $\Phi_2 \equiv 0$ and a main feature of Hadley and Ward's (1965) analysis was to test this possible result for the various fibres investigated. The isochrones for the fibres with a low molecular orientation showed this parabolic

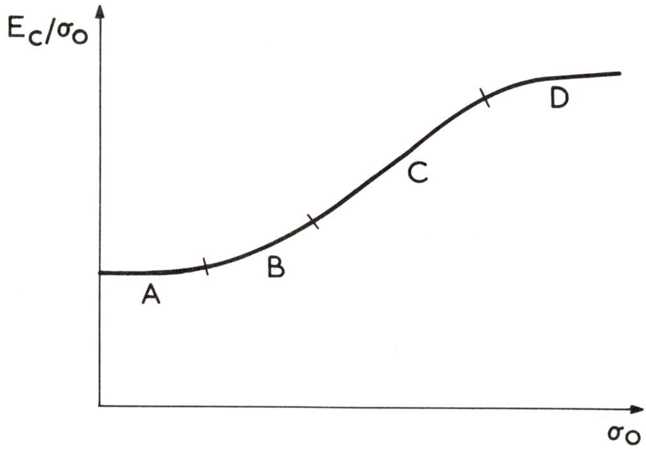

Fig. 4. Schematic diagram of various regions in the dependence of E_c/σ_0 on stress σ_0 (after Hadley and Ward, 1965).

form, but for intermediate and high molecular orientation they were straight and sigmoidal respectively. It follows that not all of the materials may be modelled by an equation in which $\Phi_2 \equiv 0$.

For the fibres of intermediate molecular orientation, the straight line isochrones can be modelled by retaining only the first two terms on the right-hand side of (5.1). However, the sigmoidal form obtained for a high molecular orientation, and shown schematically in Fig. 4, requires a higher-order polynomial if it is to be represented over the whole range: Hadley and Ward (1965) found that a fifth order relation was sufficient. If a representation is required over only a restricted range, then a simpler model will suffice. Thus in region A, E_c/σ_0 is essentially independent of σ_0 and the classical linear theory is adequate. Region B requires a parabolic form similar to that used (over the entire experimental range) to model fibres of low molecular orientation. In region C the curve is essentially linear and, in common with the fibres of intermediate orientation, the first two terms of (5.1) are sufficient. Finally, in region D the curve E_c/σ_0 is essentially independent of σ_0: behaviour unique to the fibres of high molecular orientation. Unlike the behaviour in region A, this behaviour cannot be modelled by linear theory since the curves E_r and E_c are very far from being equal which would be required to hold in the linear theory.

It should be noted that the remarks made above refer to the dominant type of behaviour exhibited by the materials considered and in the range of experimental loading. For example, the straight-line isochrones obtained for the fibres of intermediate molecular orientation imply dominant behaviour of type C with regions A and B becoming relatively small. However, if good modelling is required at the lower stress levels, this material cannot be modelled by the first two terms in (5.1) alone, although such a model is adequate over most of the experimental range. With this in mind, it can be concluded that the number of terms required to model the behaviour increases with the degree of molecular orientation. Hadley and Ward (1965) also concluded that higher order functionals become increasingly important with increasing stress level and increasing time of creep or recovery; for further details the reader is referred to the original article.

The experimental data for the fibres of high molecular orientation demonstrate the possibility that the third order relation (5.1) may not be adequate to model the behaviour of some materials. It is evident from the discussion of Section 2 that a general experimental programme for higher-order integral polynomials would certainly be prohibitive, and alternative methods of characterization must then be sought; this is treated in Chapter 9. Similar remarks apply to attempts to increase the accuracy of representation of experimental data by considering polynomials of higher order. In addition it has been shown by Gradowczyk (1969) that the accuracy of the Green-Rivlin theory does not necessarily increase as the number of terms is increased. The kernels of the multiple integrals are very sensitive to scatter in the experimental data, and the sensitivity increases as the number of terms is increased.

D. Stress-Relaxation Tests

All of the tests discussed in this section have been of creep type and have involved single- or two-step loading. Stress-relaxation tests have been performed by Lai and Findley (1968a, b), and tests involving multiple steps have been performed by several workers. However the analysis of experimental data obtained from these investigations has been made on the basis of special forms of the Green-Rivlin theory, or in the context of other constitutive relations. Discussion of these data is therefore more appropriate to the subjects of later chapters.

E. Temperature Dependence

Only a small number of experimental investigations have been concerned with the temperature-dependence of nonlinear mechanical behaviour. An investigation which has been analysed using Green-Rivlin theory is that of Morgan and Ward (1971) for the behaviour of an oriented polypropylene monofilament in the temperature range 28–60°C. Isothermal mechanical data for creep, creep and recovery, and creep with a superposed load were obtained at five temperatures. It will be recalled that the linear behaviour of thermo-rheologically-simple materials is such that material functions suffer a translation parallel to the $\log t$ axis with changing temperature. A main objective of Morgan and Ward (1971) was to determine whether a time-temperature equivalence exists in the nonlinear range.

From the results of creep tests conducted at different stress levels and different temperatures, shift factors a_T were calculated and it was found that these parameters depended only on temperature, but not on the stress level. It was shown that the form of the temperature dependence was consistent, in the relatively narrow temperature range considered, with a constant activation energy and with the Williams-Landel-Ferry relation (see Ferry, 1961). The fact that the shift factor is independent of stress level implies that the functions (5.14) obey the time-temperature equivalence with the same shift factor.

In order to study the temperature dependence of the kernel functions when their arguments are not all equal it is necessary to study the temperature-dependence of E_r and E_d. The forms of (5.25) and (5.26) suggest that an equivalence is most likely to exist if the loading time k suffers the same translation as does the variable t. For each stress level Morgan and Ward (1971) conducted a series of creep-and-recovery and creep-and-added load tests at different temperatures where the second loading occurred at the time k/a_T. The values of a_T were those determined from the previous creep tests, and this choice of time implies that the creep strain is the same at all temperatures immediately before the load is changed. Under these conditions it was found that E_r and E_d also obeyed time-temperature equivalence under the same shift factors a_T that were determined for creep behaviour.

Since E_c, E_r and E_d satisfied time-temperature equivalence under the same set of shift factors a_T which are independent of stress level, Morgan and Ward (1971) concluded that the material functions Φ_n also satisfied this relation. They further interpreted this conclusion as an indication that the multiple integral representation has a physical significance and that the three kernels are associated with the same molecular processes.

Before leaving the work of Morgan and Ward (1971) it is of interest to examine the form of the nonlinear behaviour exhibited by their data. Isochronous curves of E_c/σ_0 versus σ_0 approximate to region C in Fig. 4. with gradients which decreased slightly with increasing σ_0. The corresponding forms of the material functions Φ_n may be described with reference to the notation (5.22) and (5.23); Fig. 5 shows the forms of Φ_n^* as functions of t for the value $\bar{\sigma} = 10^9$ dyne/cm^2. It will be noted that Φ_1^* and Φ_2^* are positive, whereas Φ_3^* is negative; this is in contrast to the behaviour of polyethylene (Lifshitz and Kolsky, 1967; see Section 3A) for which Φ_1^* and Φ_3^* are positive and Φ_2^* is negative. Thus the signs of the higher order creep functions do not obey a universal rule, although available experimental data point to the fact that their magnitudes increase

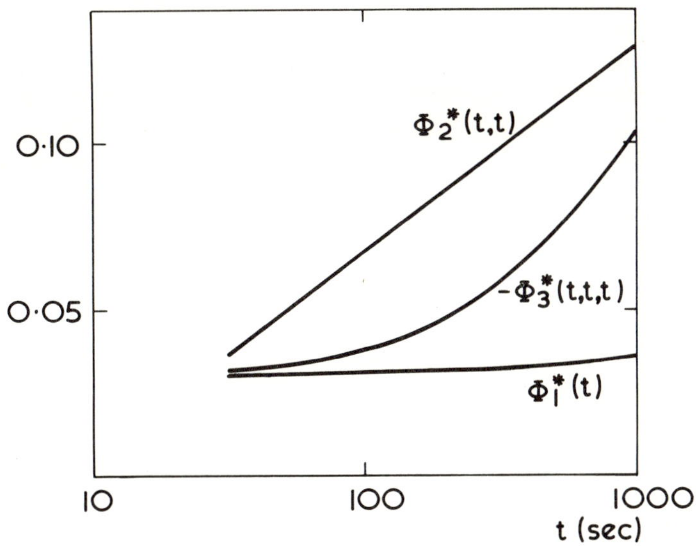

Fig. 5. Creep functions for oriented polypropylene (after Morgan and Ward, 1971).

with increasing time. Since the highest stress level considered by Morgan and Ward (1971) is equivalent to $\sigma_* = 0.5$ it is evident from Fig. 5 that their data involves a significant degree of nonlinearity.

4. DETERMINATION OF COMPLEX MODULI THROUGH OSCILLATORY TESTS

In principle the nonlinear behaviour of a material may be determined through oscillatory tests which generalize the tests described in Chapter 3, Section 1C for a linear material. The theory of these tests has been considered by Nakada (1960) and Lockett and Gurtin (1964) but experimental data appears to be lacking at the present time. In part this is due to the large number of experiments which are required for adequate material characterization, but it may also be due to recognition of a possible practical restriction. Forced vibration of a viscoelastic material is accompanied by dynamic heating of the specimen and, in the loading range required to exhibit significant nonlinearity, this may be sufficient to prevent adequate temperature control. Since the mechanical behaviour of the materials considered shows a marked temperature dependence, and since the Green-Rivlin theory assumes isothermal conditions, the proposed oscillatory tests may be of only limited practical value. Nevertheless the method may be suitable for some materials, and an outline of the method is therefore of interest.

The theory and notation of Section 5, Chapter 4 is used below to discuss one-dimensional tests which lead to values for the complex moduli of a material. Determination of complex compliances can be treated by interchanging the roles of stress and strain in the analysis which follows. For the sake of brevity the analysis is presented for a second-order constitutive relation: the third-order theory has been treated by Lockett and Gurtin, but it is more complicated. It will be recalled from the discussion of Section 3C that a second-order equation is adequate for some materials.

Consider a homogeneous oscillatory strain history which has a finite number of discrete frequency components:

$$\epsilon(t) = \sum_{m=1}^{M} A_m \sin \omega_m t \tag{5.31}$$

where A_m are constant amplitudes and ω_m are the discrete frequencies. Upon substitution of (5.31) into (4.53) with $n = 1$ it is found that

$$\begin{aligned} \sigma_1(t) &= \sum_{m=1}^{M} A_m \omega_m \int_0^\infty \Phi_1(s) \cos [\omega_m(t-s)]\, ds \\ &= \sum_{m=1}^{M} A_m [E_{1i}(\omega_m) \cos \omega_m t + E_{1r}(\omega_m) \sin \omega_m t] \end{aligned} \tag{5.32}$$

where use has been made of (4.54) and (4.55), and E_{1r} and E_{1i} denote the real and imaginary parts of the first-order complex modulus E_1. In a similar manner

$$\sigma_2(t) = \sum_{m=1}^{M} \sum_{p=1}^{M} A_m A_p \omega_m \omega_p \iint_0^\infty \Phi_2(s_1, s_2) \cos [\omega_m(t-s_1)] \cos [\omega_p(t-s_2)]\, ds_{(2)} \tag{5.33}$$

and, using (4.54), (4.55) and standard trigonometric identities, this can be written in the form

$$\sigma_2(t) = \tfrac{1}{2} \sum_{m=1}^{M} \sum_{p=1}^{M} A_m A_p[-E_{2r}(\omega_m, \omega_p) \cos [(\omega_m + \omega_p) t]$$
$$+ E_{2i}(\omega_m, \omega_p) \sin [(\omega_m + \omega_p) t] + E_{2r}(\omega_m, -\omega_p) \cos [(\omega_m - \omega_p) t]$$
$$- E_{2i}(\omega_m, -\omega_p) \sin [(\omega_m - \omega_p) t]]. \tag{5.34}$$

Calculation of $\sigma_n(t)$ for $n > 2$ is straightforward but tedious. It will now be shown how the complex moduli may be determined in experiments where the applied strain has the form (5.31).

A. Applied Strain with Single Frequency Component

Consider the special case for which $M = 1$ and write $A_1 = A$, $\omega_1 = \omega$ for convenience. The stress response to this applied strain, which will be denoted by $V(\omega, A; t)$, is given by (4.52), (5.32) and (5.34) to be

$$\sigma(t) \equiv V(\omega, A; t) = X(\omega, A) + \sum_{n=1}^{2} [Y_n(\omega, A) \cos n\omega t + Z_n(\omega, A) \sin n\omega t] \tag{5.35}$$

where the second-order theory is considered and

$$X(\omega, A) = \tfrac{1}{2}A^2 E_{2r}(\omega, -\omega) \qquad Y_1(\omega, A) = AE_{1i}(\omega)$$
$$Y_2(\omega, A) = -\tfrac{1}{2}A^2 E_{2r}(\omega, \omega) \qquad Z_1(\omega, A) = AE_{1r}(\omega)$$
$$Z_2(\omega, A) = \tfrac{1}{2}A^2 E_{2i}(\omega, \omega). \tag{5.36}$$

If $E_2 \equiv 0$, equations (5.35) and (5.36) reduce to the linear form discussed in Chapter 3, Section 1C.

When a spectral analysis is made of the stress $\sigma(t)$ required to support the applied strain, equations (5.35) and (5.36) show that the frequency components yield values for

$$E_{1r}(\omega), \qquad E_{1i}(\omega), \qquad E_{2r}(\omega, \omega), \qquad E_{2i}(\omega, \omega), \qquad E_{2r}(\omega, -\omega) \tag{5.37}$$

at the particular frequency ω. To express these moduli as functions of frequency, the whole procedure must be repeated for a range of values of ω; it will be assumed that N values are sufficient.

The set of quantities (5.37) contains one member evaluated at a negative value of its argument; this results from an experiment conducted at the positive frequency ω. The negative value does not imply a negative frequency, but is merely a consequence of the transform relation. By making use of (4.54),

(4.55) and the assumed symmetry of Φ_2, it is readily shown that values of the moduli for other negative arguments are given by

$$E_{1r}(-\omega) = E_{1r}(\omega), \qquad E_{1i}(-\omega) = -E_{1i}(\omega)$$
$$E_{2r}(-\omega, -\omega) = E_{2r}(\omega, \omega), \qquad E_{2i}(-\omega, -\omega) = -E_{2i}(\omega, \omega)$$
$$E_{2i}(\omega, -\omega) \equiv 0. \tag{5.38}$$

B. Applied Strain with Two Frequency Components

It is obvious from the previous discussion that it is not possible to determine completely the functions $E_{2r}(\omega_1, \omega_2)$, $E_{2i}(\omega_1, \omega_2)$ from tests involving a single frequency component. In order to determine these functions when their arguments have different magnitudes consider an applied strain with $M = 2$. The stress response is then given by

$$\sigma(t) = \sum_{m=1}^{2} V(\omega_m, A_m; t) + I(\omega_1, \omega_2, A_1, A_2; t) \tag{5.39}$$

where

$$\begin{aligned} I(\omega_1, \omega_2, A_1, A_2; t) \\ = A_1 A_2[&-E_{2r}(\omega_1, \omega_2) \cos [(\omega_1 + \omega_2) t] + E_{2i}(\omega_1, \omega_2) \sin [(\omega_1 + \omega_2) t] \\ &+ E_{2r}(\omega_1, -\omega_2) \cos [(\omega_1 - \omega_2) t] - E_{2i}(\omega_1, -\omega_2) \sin [(\omega_1 - \omega_2) t]]. \end{aligned} \tag{5.40}$$

The first terms on the right of (5.39) are known from the single-frequency tests, so that the function I can be determined from the experimental data. A spectral analysis of this function then yields the values of the modulus functions appearing in (5.40). For complete characterization the procedure must be repeated for other pairs of values (ω_1, ω_2); due to symmetry of material functions it is sufficient to consider $\omega_1 < \omega_2$ and the number of tests is $\frac{1}{2}N(N-1)$. The total number of tests for determining the moduli in a second-order constitutive relation is $\frac{1}{2}N(N+1)$ and this may be compared with the number $N+2$ which is the number of creep or stress-relaxation tests required by a second-order theory. The reason for the larger requirement in oscillatory tests is that each of these experiments provides less information than a single creep or stress-relaxation test. The former determines values of moduli at discrete points in the argument space but the latter provide information along lines in the argument space. These conclusions extend those for linear theory where a single creep test is sufficient for one-dimensional characterization, but a set of oscillatory tests must be performed.

Chapter 6 Three-Dimensional Green–Rivlin Behaviour

In Chapter 5 it was shown that choice of a third-order Green-Rivlin expression to characterize one-dimensional nonlinear behaviour requires a considerable experimental programme for determining the appropriate material parameters. Although the tests are all of the multiple-step type, and can be performed on standard testing equipment, many individual tests are needed. A similar approach to three-dimensional isotropic behaviour can be expected to result in an even greater requirement which, in its most general form, would be prohibitive. This prediction is confirmed in Section 1 where an experimental programme for the third-order three-dimensional theory is considered. This is not proposed as a routine testing procedure, but it demonstrates the type and number of experiments that would be required if the full theory were used. In addition it demonstrates the futility of mathematical theories which do not consider the consequent experimental requirements. On the positive side, the programme provides a framework within which it is possible to analyse experimental data for particular situations in the context of their significance to a general description of material properties.

It is shown in Section 1 that a complete experimental investigation involves a large number of experiments. Although these are of a few basic types, some of the experimental requirements are difficult to realize and this reduces the practical significance of the general analysis. In an attempt to provide a basis for feasible experimental programmes, the remainder of the chapter is devoted to special theories which describe incompressible materials either under general deformations or under conditions of plane stress and plane strain.

1. THIRD-ORDER THREE-DIMENSIONAL BEHAVIOUR

In this section an experimental programme is given for determining the material parameters in a three-dimensional theory in which terms up to and including those of third order in the input matrix are retained. The relevant constitutive equation is (4.28) in which **P** and **Q** can be given either of the interpretations (4.26) or (4.27). In (4.28) ψ_1 and ψ_2 are functions of $t-\tau_1$; $\psi_3, \ldots, \psi_6$ are functions of $t-\tau_1$ and $t-\tau_2$; $\psi_7, \ldots, \psi_{12}$ are functions of $t-\tau_1$, $t-\tau_2$ and $t-\tau_3$. Thus an experimental programme is required to determine two functions of a single variable, four functions of two variables, and six functions of three variables. In the present discussion only a summary of the necessary experimental programme is given; motivation for the experiments, and the analysis for determining material functions from experimental data, can be found in the original paper by Lockett (1965).

Since the matrices **P** and **Q** may be given the interpretations (4.26) and (4.27), the experiments may be of either the creep or stress-relaxation type. In general (4.28) is an implicit stress/strain relation due to the appearance of the matrix **R** in the definition of **P** or **Q**; this is an additional complication in application of (4.28). It is therefore natural to consider first situations in which

Table 1. Experimental programme for determining $\psi_1, \ldots, \psi_{11}$ (after Lockett, 1965)

	Input	Number of experiments	Outputs to be measured
1	$P_{11} = A_1 H(t)$	1	Q_{11}, Q_{33}
2	$P_{11} = A_2 H(t)$	1	
3	$P_{11} = A_3 H(t)$	1	
4	$P_{11} = A_1 H(t) + B_1 H(t-k)$	N	
5	$P_{11} = A_1 H(t) + B_2 H(t-k)$	N	
6	$P_{11} = A_2 H(t) + B_1 H(t-k)$	N	
7	$P_{11} = AH(t) + BH(t-k) + CH(t-l)$	$\frac{1}{2}N(N-1)$	
8	$P_{11} = A_1 H(t),\ P_{22} = B_1 H(t-k)$	N	Q_{11}, Q_{22}, Q_{33}
9	$P_{11} = A_1 H(t),\ P_{22} = B_2 H(t-k)$	N	
10	$P_{11} = A_2 H(t),\ P_{22} = B_1 H(t-k)$	N	
11	$P_{11} = AH(t),\ P_{22} = BH(t-k) + CH(t-l)$	$\frac{1}{2}N(N-1)$	
12	$P_{11} = AH(t-k),\ P_{22} = BH(t) + CH(t-l)$	N^2	
13	$P_{11} = AH(t),\ P_{22} = BH(t-k),\ P_{33} = CH(t-l)$	$\frac{1}{2}N(N-1)$	

$\mathbf{R} = \mathbf{I}$, so that **P** and **Q** merely represent $\boldsymbol{\sigma}$ and **E**. These situations arise when $P_{ij} = 0$ for $i \neq j$, and it has been shown (Lockett, 1965) that all of the material functions except ψ_{12} may be determined from experiments of this form.

The complete set of homogeneous input programmes required to determine $\psi_1, \ldots, \psi_{11}$ is summarized in Table 1. In these experiments the magnitudes A, B and C are arbitrary if given without suffixes. Equality of magnitudes which are given with suffixes need only be taken within the same group of tests (groups of tests are separated by horizontal lines).

To determine the material function ψ_{12} it is necessary to perform additional tests for which there is at least one off-diagonal component in the matrix **P**. The function ψ_{12} may be determined via the experiments given in Table 2.

In all the experiments listed in this section, k and l take only positive values. If it is sufficient to consider $k < l$, then this fact is indicated by the statement that only $\frac{1}{2}N(N-1)$ experiments of the particular type need be performed. If the number of experiments is shown as N^2, then both of the cases $k < l$ and $k > l$ should be considered.

All of the experiments required for three-dimensional characterization involve multiple-step inputs and are similar to those discussed in Chapter 5, Section 2 for the one-dimensional theory. However, in the three-dimensional theory, it is often necessary to apply different steps of the input programme to different components of the input matrix **P** and to take measurements of more than one component of the output matrix **Q**. Thus, for example, experiments 1–7 differ from those required by the one-dimensional theory (see Table 1 of Chapter 5) only in so far as an extra, transverse, output Q_{33} has to be measured. This is equivalent to a direct measurement of Poisson's ratio in determining the three-dimensional properties of a linear isotropic material.

Table 2. Additional experiments for determining ψ_{12} (after Lockett, 1965)

Input	Number of experiments	Outputs to be measured
14 $P_{11} = AH(t),\ P_{12} = P_{21} = BH(t-k)$	N	Q_{22}
15 $P_{11} = AH(t-k),\ P_{12} = P_{21} = BH(t)$	N	
16 $P_{11} = AH(t),\ P_{12} = P_{21} = BH(t-k) + CH(t-l)$	$\frac{1}{2}N(N-1)$	
17 $P_{11} = AH(t-k),\ P_{12} = P_{21} = BH(t) + CH(t-l)$	N^2	

Experiments 8–12 are biaxial tests in which inputs of stress or strain are applied at various times. Biaxial tests on sheets of material can be performed successfully (see, for example, McGuirt and Lianis, 1967b) and this requirement is practically feasible. However, it is necessary to measure all of the non-zero components of the output matrix (Q_{11}, Q_{22} and Q_{33}). Whereas this is possible for components in the plane of the test, measurement of the thickness component Q_{33} in a biaxial sheet test may not be practicable. Experiment 13, which is a triaxial test, also requires measurement of the three components of **Q**. Triaxial tests in which inputs are applied mechanically in all three directions involve insuperable experimental difficulties. However, it may be possible to apply triaxial loads in a creep test by performing biaxial tests under a suitably changing hydrostatic pressure although the author is not aware of any successful experimental investigations of this type.

Finally, experiments 14–17 involve the combined effect of a normal input and a shear input. In principle this can be achieved in an experiment such as the combined tension and torsion of a hollow cylinder. However, since $\mathbf{R} \neq \mathbf{I}$, it is necessary either to apply an input which has the required time variation in a coordinate system which rotates with the deformation, or to compute outputs in such a coordinate system. This fact makes the conditions of experiments 14–17 very difficult to achieve exactly. In special circumstances the requirement on coordinate systems may be relaxed [see, for example, the work of Lifshitz and Kolsky (1967) which was discussed in Chapter 5, Section 3] but, in general, this is not possible.

The total number of experiments proposed in Tables 1 and 2 is $4N^2 + 6N + 3$ and it should be remembered that this is the minimum requirement and does not allow for repetition of individual experiments in order to improve accuracy through averaging of data. For the adequate value $N = 10$ the number of experiments is 463 and this compares with the number 78 for the corresponding one-dimensional theory. Thus the increase in complexity introduced by considering general three-dimensional behaviour is considerable, and the magnitude of the experimental task makes equation (4.28) unsuitable for characterizing even a single material on which it might be decided to perform a special investigation. Thus, as stated already, the discussion given above is useful only as a framework within which to consider the results of individual experiments.

Although the general equation (4.28) is unsuitable for practical use, it is possible that certain special cases could lead to a feasible experimental programme, although this is unlikely to yield a simple programme: this is indicated by consideration of the one-dimensional problem in Chapter 5, Section 2. The remainder of this chapter is devoted to the simplifications obtained in two special cases; one involving a restriction on the type of material behaviour, the other concerned with a particular class of loading situations.

2. INCOMPRESSIBLE MATERIALS: STRESS-RELAXATION FORMULATION

A. Derivation of Constitutive Relation

A constitutive relation for an incompressible Green-Rivlin material has been obtained by Pipkin (1964), but the derivation requires some comment. Consider the formulation (4.28) in which

$$\mathbf{Q} = \mathbf{R}^T \boldsymbol{\sigma} \mathbf{R}, \qquad \mathbf{P} = \mathbf{E} \tag{6.1}$$

so that the ψ_α are stress-relaxation functions. The fact that no volume changes can occur in an incompressible material requires that

$$\det |2\mathbf{E} + \mathbf{I}| = 1 \tag{6.2}$$

which, upon expansion of the determinant, can be written as

$$\operatorname{tr} \mathbf{E} = \operatorname{tr} \mathbf{E}^2 - (\operatorname{tr} \mathbf{E})^2 - \tfrac{2}{3}[2 \operatorname{tr} \mathbf{E}^3 - 3 \operatorname{tr} \mathbf{E} \operatorname{tr} \mathbf{E}^2 + (\operatorname{tr} \mathbf{E})^3]. \tag{6.3}$$

It is now supposed that $\mathbf{E}$ is small in some sense and the theory is considered in which terms are retained up to and including those of third order in $\mathbf{E}$. By re-substitution for $\operatorname{tr}\mathbf{E}$ on the right-hand side of (6.3) it follows that

$$\operatorname{tr}\mathbf{E} = \operatorname{tr}\mathbf{E}^2 - \tfrac{4}{3}\operatorname{tr}\mathbf{E}^3 \tag{6.4}$$

to third order in $\mathbf{E}$. From this it follows that wherever $T_\alpha = \operatorname{tr}\dot{\mathbf{E}}(\tau_\alpha)$ appears in (4.28), it can be replaced by higher-order terms. There is consequently no loss in generality in omitting these terms in (4.28) so that

$$\mathbf{Q}(t) = \int_{-\infty}^{t} \psi_2 \mathbf{M}_1 \, d\tau_{(1)} + \iint_{-\infty}^{t} [\mathbf{I}\psi_4 T_{12} + \psi_6 \mathbf{M}_1 \mathbf{M}_2] \, d\tau_{(2)} + \iiint_{-\infty}^{t} [\mathbf{I}\psi_7 T_{123} + \psi_{10} T_{12} \mathbf{M}_3 + \psi_{12} \mathbf{M}_1 \mathbf{M}_2 \mathbf{M}_3] \, d\tau_{(3)} \tag{6.5}$$

to third-order, and where the ψ_α are in general different from those appearing in (4.28).

In addition to the constraint of no volume change, Pipkin (1964) assumes that as a reaction to this constraint, there arises a hydrostatic pressure which is arbitrary in so far as the constitutive relation is concerned. This implies that when the deformation history is known the stress is undetermined with respect to an arbitrary hydrostatic pressure. If this assumption is valid an arbitrary pressure term must be included in (6.5), and the terms involving ψ_4 and ψ_7 can be included in this pressure. Equation (6.5) then has the form

$$\mathbf{Q}(t) = \int_{-\infty}^{t} \psi_2 \mathbf{M}_1 \, d\tau_{(1)} + \iint_{-\infty}^{t} \psi_6 \mathbf{M}_1 \mathbf{M}_2 \, d\tau_{(2)} + \iiint_{-\infty}^{t} [\psi_{10} T_{12} \mathbf{M}_3 + \psi_{12} \mathbf{M}_1 \mathbf{M}_2 \mathbf{M}_3] \, d\tau_{(3)} \tag{6.6}$$

where $\mathbf{Q}$ is to be interpreted as

$$\mathbf{Q} = \mathbf{R}^T \mathbf{s} \mathbf{R} + p\mathbf{I} \tag{6.7}$$

$\mathbf{s}$ is the deviatoric stress matrix $\mathbf{s} = \boldsymbol{\sigma} - \tfrac{1}{3}\mathbf{I}\operatorname{tr}\boldsymbol{\sigma}$, and p is a hydrostatic pressure which, in general, is not equal to $\tfrac{1}{3}\operatorname{tr}\boldsymbol{\sigma}$.

It is necessary to make some comments concerning the relevance of Pipkin's second assumption to the definition of an incompressible material. If the relation between stress and strain history is defined by the constitutive relation only to within an arbitrary hydrostatic pressure, then it follows that the same strain history can be supported by an infinite number of stress fields which differ by hydrostatic contributions. In particular, pure shear deformation of the material requires the same shear loading no matter what hydrostatic pressure is superimposed. Now it is not obvious that this is a condition which can be expected to apply to real materials and, in the opinion of the author, this is not a condition that naturally forms part of the definition of an incompressible material. Any dependence of shear response on hydrostatic loading is an interaction effect which can be expected to occur only in a nonlinear theory. Thus these comments do not apply to the linear behaviour discussed earlier and Pipkin's second assumption, which is accepted without question in the linear theory, is valid there.

In the remainder of this chapter the constitutive relation for an incompressible material will be taken as (6.6) with **Q** defined by (6.7). It is emphasized that this equation depends on the assumption regarding the effect of hydrostatic pressures in addition to the requirement of no volume change. If a less restrictive definition of incompressibility is required which includes only the latter assumption, then (6.5) and (6.1) must be used and a correspondingly larger experimental programme is necessary in order to determine the additional functions ψ_4 and ψ_7.

When formulating an experimental programme for incompressible materials Lockett (1965) considered the special case $p \equiv 0$ for which the relevant components of **Q** are more readily determined from experimental data. In general $p \neq 0$ and a modification of Lockett's (1965) analysis is required in order to eliminate the unknown quantity p and this requires measurement of an additional component of the stress matrix. This more general analysis is given here; an outline of the experimental programme follows in the next sub-section and the motivation for the experiments is discussed in Section 2C, together with comments on the manner in which data are to be used to determine material functions.

B. Experimental Programme for Determining Stress-Relaxation Functions

Although the constitutive relation (6.6) is a special case of the more general relation (4.28), the programme of experiments required for an incompressible material cannot be taken directly from those discussed in Section 1 for a compressible material. This is due to the fact that it is not possible, for example, for E_{11} to be the only non-zero component of strain in a deformed incompressible material. Some other strain component must also be non-zero in order to preserve the volume of material elements. In the experiments discussed below it is assumed that certain strain components are *imposed* on the material, whereas the component E_{33} is *allowed* to take whatever value is required by the incompressibility condition (6.2). Thus, in particular, E_{33} will have a step at each instant of time at which at least one of the imposed strain components has a step. The complete set of experiments required is shown in Table 3.

Experiments 7, 8, 10 need be considered only for values of k and l satisfying $0 < k < l$; in experiment 11 both of the cases $k < l$ and $k > l$ must be considered.

In experiments 1–8 the imposed strains are such that $\mathbf{R} = \mathbf{I}$ and stresses are measured in a fixed coordinate system; this is a great simplification. The tests are essentially biaxial in nature, since $E_{22} \equiv 0$ must be imposed on the material specimen in addition to the strains E_{11} given in Table 3. These experiments can be performed in a biaxial test on a sheet of material where the plane of the sheet is parallel to the (X_1, X_2)-plane and the sides of the sheet are parallel to

Table 3. Experimental programme for incompressible material (after Lockett, 1965)

Imposed strain input*	Number of experiments
1 $E_{11} = A_1 H(t)$	1
2 $E_{11} = A_2 H(t)$	1
3 $E_{11} = A_1 H(t) + B_1 H(t-k)$	N
4 $E_{11} = A_2 H(t) + B_2 H(t-k)$	N
5 $E_{11} = A_3 H(t) + B_3 H(t-k)$	N
6 $E_{11} = A_4 H(t) + B_4 H(t-k)$	N
7 $E_{11} = A_1 H(t) + B_1 H(t-k) + C_1 H(t-l)$	$\frac{1}{2}N(N-1)$
8 $E_{11} = A_2 H(t) + B_2 H(t-k) + C_2 H(t-l)$	$\frac{1}{2}N(N-1)$
9 $E_{11} = AH(t),\ E_{12} = E_{21} = BH(t-k)$	N
10 $E_{11} = AH(t),\ E_{12} = E_{21} = BH(t-k) + CH(t-l)$	$\frac{1}{2}N(N-1)$
11 $E_{11} = AH(t) + BH(t-k),\ E_{12} = E_{21} = CH(t-l)$	N^2

* In addition to the imposed strain inputs the component E_{33} is allowed to take whatever value is required by the incompressibility condition. All other strain components are held at zero.

the coordinate axes. Since E_{33} is allowed to take whatever value is required by the condition of zero volume change, it is possible to set $\sigma_{33} = 0$ and to measure σ_{11} and σ_{22} in order to determine $\mathbf{s}$.

Experiments 9–11 require the combined effects of extension and shear and may not be easy to achieve in practice. If extension and torsion of a tube is considered, where axes are chosen at a typical point in the material so that X_1 is parallel to the tube axis and X_3 is in the thickness direction, then extension and torsion can provide the required values for E_{11} and E_{12}. Since E_{33} is allowed to take the value required by the condition of zero volume change, no constraint is imposed on the material in the thickness direction. However, although it is possible to constrain the specimen so that $E_{22} = 0$ at the ends of the tube, it is not evident that this will result in a homogeneous deformation in which this condition is satisfied away from the ends. In any case, since $\mathbf{R} \neq \mathbf{I}$, the outputs $\mathbf{Q}$ must be calculated from measurements of all of the components of $\boldsymbol{\sigma}$ together with the form of $\mathbf{R}$ appropriate to the imposed strains $\mathbf{E}$.

The total number of experiments given in Table 3 is $\frac{5}{2}N^2 + \frac{7}{2}N + 2$ which implies 287 experiments when $N = 10$. This is a relatively modest reduction when compared with the total of 463 required for a compressible material, and still represents a prohibitive requirement.

C. Analysis of Experimental Data

In this subsection the motivation behind the experimental programme is given, together with an indication of the method of data analysis. In experiments 1–8 the only non-zero components of $\mathbf{E}$ are of the form $E_{11} = f(t)$, $E_{33} = h(t)$, so that (6.6) gives

$$Q_{11} = \psi_2 \dot{f}_1 + \psi_6 \dot{f}_1 \dot{f}_2 + (\psi_{10} + \psi_{12}) \dot{f}_1 \dot{f}_2 \dot{f}_3 + \psi_{10} \dot{h}_1 \dot{h}_2 \dot{f}_3$$

$$Q_{33} = \psi_2 \dot{h}_1 + \psi_6 \dot{h}_1 \dot{h}_2 + (\psi_{10} + \psi_{12}) \dot{h}_1 \dot{h}_2 \dot{h}_3 + \psi_{10} \dot{f}_1 \dot{f}_2 \dot{h}_3 \tag{6.8}$$

and $Q_{22} \equiv 0$. In (6.8) $\psi_2 \dot{f}_1$ is an abbreviated notation for

$\int_{-\infty}^{t} \psi_2(t - \tau_1) \dot{f}(\tau_1)\, d\tau_1$, etc.

Since $\mathbf{R} = \mathbf{I}$ the result for Q_{22} gives $p = -s_{22}$ so that p, and hence Q_{11} and Q_{33}, become measurable quantities.

In experiments 1 and 2, f and h are given by

$$f(t) = AH(t), \qquad h(t) = XH(t), \qquad X = -A/(1 + 2A) \tag{6.9}$$

where (6.2) has been used. Equations (6.8) and (6.9) can now be used to write

$$(A^{-1} Q_{11} - X^{-1} Q_{33})(A - X)^{-1} = \psi_6(t, t) + (A + X)\psi_{12}(t, t, t) \tag{6.10}$$

$$X^{-1} Q_{33} - X\psi_6(t, t) - X^2\psi_{12}(t, t, t) = \psi_2(t) + (A^2 + X^2)\psi_{10}(t, t, t). \tag{6.11}$$

Thus experiments 1–2 and equation (6.10) yield values for

$$\psi_6(t, t), \qquad \psi_{12}(t, t, t) \tag{6.12}$$

and equation (6.11) then determines

$$\psi_2(t), \qquad \psi_{10}(t, t, t) \tag{6.13}$$

The double-step tests 3–6 are of the form

$$f = AH(t) + BH(t - k), \tag{6.14}$$

which implies that

$$h = XH(t) + YH(t - k) \tag{6.15}$$

where X and Y are given by (6.2), (6.14) and (6.15). The expression for Q_{11} is

$$\begin{aligned} Q_{11} = 2AB\psi_6(t, t-k) &+ A^2 B[\psi_{10}(t, t, t-k) + 2\psi_{10}(t, t-k, t) \\ &+ \psi_{12}(t, t-k, t) + 2\psi_{12}(t, t, t-k)] + AB^2[\psi_{10}(t-k, t-k, t) \\ &+ 2\psi_{10}(t, t-k, t-k) + \psi_{12}(t-k, t, t-k) + 2\psi_{12}(t, t-k, t-k)] \\ &+ X^2 B\psi_{10}(t, t, t-k) + Y^2 A\psi_{10}(t-k, t-k, t) \\ &+ 2XYA\psi_{10}(t, t-k, t) + 2XYB\psi_{10}(t, t-k, t-k) + \cdots \end{aligned} \tag{6.16}$$

where known functions have not been written explicitly and where use has been made of the symmetries (4.30). The expression for Q_{33} is obtained from that for Q_{11} if X and Y are interchanged with A and B respectively. Since Q_{11} and Q_{33} are measurable quantities, experiments 3–6 provide sufficient equations for the determination of

$$\psi_6(t, t-k) \tag{6.17}$$

$$\psi_{10}(t, t, t-k), \quad \psi_{10}(t-k, t-k, t), \quad \psi_{10}(t, t-k, t), \quad \psi_{10}(t, t-k, t-k) \tag{6.18}$$

$$\psi_{12}(t, t-k, t) + 2\psi_{12}(t, t, t-k), \quad \psi_{12}(t-k, t, t-k) + 2\psi_{12}(t, t-k, t-k) \tag{6.19}$$

When these experiments and calculations are repeated for a range of values of k, ψ_6 is determined completely through (6.17); (6.18) yields the values of ψ_{10} when any two of its arguments are equal. Expressions (6.19) provide some intermediate information in the determination of ψ_{12}.

For tests 7–8 the strain components are such that

$$\begin{aligned} f &= AH(t) + BH(t-k) + CH(t-l), \\ h &= XH(t) + YH(t-k) + ZH(t-l) \end{aligned} \tag{6.20}$$

where X, Y and Z may be found from (6.2). The expression for the measurable quantity Q_{11} is now

$$\begin{aligned} Q_{11} = 2ABC[\psi_{10}(t, t-k, t-l) &+ \psi_{10}(t, t-l, t-k) + \psi_{10}(t-k, t-l, t) \\ &+ \psi_{12}(t, t-k, t-l) + \psi_{12}(t-k, t, t-l) + \psi_{12}(t, t-l, t-k)] \\ &+ 2XYC\psi_{10}(t, t-k, t-l) + 2XZB\psi_{10}(t, t-l, t-k) \\ &+ 2YZA\psi_{10}(t-k, t-l, t) + \cdots \end{aligned} \tag{6.21}$$

where known terms have not been written explicitly. The expression for Q_{33} is similar to (6.21) but has the roles of A, B, C and X, Y, Z interchanged.

Experiments 7–8 of Table 3 and the expressions for Q_{11} and Q_{33} now allow determination of

$$\psi_{10}(t, t-k, t-l), \qquad \psi_{10}(t, t-l, t-k), \qquad \psi_{10}(t-k, t-l, t) \tag{6.22}$$

$$\psi_{12}(t, t-k, t-l) + \psi_{12}(t-k, t, t-l) + \psi_{12}(t, t-l, t-k). \tag{6.23}$$

When these experiments are repeated for a range of values of k and l the function ψ_{10} is determined completely. Expression (6.23) provides further intermediate information about ψ_{12}.

To complete the determination of ψ_{12} it is necessary to consider situations in which the only non-zero components of $\mathbf{E}$ are $E_{11} = f(t)$, $E_{12} = E_{21} = a(t)$, $E_{33} = h(t)$ so that $\mathbf{R} \neq \mathbf{I}$ and (6.6) yields

$$Q_{11} = \psi_{12}\dot{f}_1\dot{f}_2\dot{f}_3 + 2\psi_{12}\dot{f}_1\dot{a}_2\dot{a}_3 + \cdots, \qquad Q_{22} = \psi_{12}\dot{a}_1\dot{f}_2\dot{a}_3 + \cdots$$

$$Q_{33} = \psi_{12}\dot{h}_1\dot{h}_2\dot{h}_3 + \cdots, \qquad Q_{12} = \psi_{12}\dot{f}_1\dot{f}_2\dot{a}_3 + \psi_{12}\dot{a}_1\dot{a}_2\dot{a}_3 + \cdots \tag{6.24}$$

where an abbreviated notation similar to that used in (6.8) has been employed, and known functions have not been written explicitly. For the strain input of experiment 9 these expressions yield

$$Q_{11} - Q_{22} = AB^2[2\psi_{12}(t, t-k, t-k) - \psi_{12}(t-k, t, t-k)] + \cdots$$

$$Q_{12} = A^2 B\psi_{12}(t, t, t-k) + \cdots \tag{6.25}$$

Since the left-hand sides of these equations are measurable quantities the experiment determines

$$\psi_{12}(t, t, t-k), \qquad 2\psi_{12}(t, t-k, t-k) - \psi_{12}(t-k, t, t-k). \tag{6.26}$$

When this experiment and calculation are repeated for a range of values of k, (6.19) and (6.26) completely determine ψ_{12} when any two of its arguments are equal.

Experiment 10 now yields the relation

$$Q_{11} - Q_{22} = 2ABC[\psi_{12}(t, t-k, t-l) + \psi_{12}(t, t-l, t-k) - \psi_{12}(t-k, t, t-l)] + \cdots \tag{6.27}$$

so that the expression within square brackets can be found; in conjunction with (6.23) this permits calculation of

$$\psi_{12}(t-k, t, t-l) \tag{6.28}$$

The expressions for $Q_{22} - Q_{33}$ and Q_{12} do not contain any new information. Finally, experiment 11 provides

$$Q_{12} = ABC[\psi_{12}(t, t-k, t-l) + \psi_{12}(t-k, t, t-l)] + \cdots \tag{6.29}$$

and it is possible to determine $\psi_{12}(t,\ t-k,\ t-l)$ from (6.28) and (6.29); $\psi_{12}(t, t-l, t-k)$ now follows from (6.23). After repetition for a range of values of k and l, ψ_{12} is determined completely; this completes the determination of material functions.

3. INCOMPRESSIBLE MATERIALS: CREEP FORMULATION

A. Derivation of Constitutive Relation

In this section the alternative form of constitutive relation for an incompressible material is considered in which **P** and **Q** initially have the definitions (4.27); the analysis is due to Lockett and Stafford (1969). In keeping with the derivation of (6.6) it will be assumed that the material is unaffected by hydrostatic pressure, and it follows that (4.27) may be replaced by

$$\mathbf{Q} = \mathbf{E}, \qquad \mathbf{P} = \mathbf{R}^T \mathbf{s}\mathbf{R} + p\mathbf{I} \tag{6.30}$$

where $\mathbf{s}$ is the deviatoric stress matrix. Since the stress is defined by the constitutive relation only to within an arbitrary hydrostatic pressure, $p \neq \frac{1}{3}\operatorname{tr}\boldsymbol{\sigma}$. The assumption that the deformation is unaffected by hydrostatic pressures implies that terms involving p should give no contribution to $\mathbf{E}$ no matter what value is assigned to $\mathbf{s}$. This condition leads to a number of equations relating the functions ψ_α which are creep functions in this formulation and are different from the stress-relaxation functions considered in Section 2. In particular it follows that

$$6\psi_7 = \tfrac{1}{3}\psi_{12}^* - \psi_{10}^*. \tag{6.31}$$

In this relation $\psi_7(s_1, s_2, s_3)$ is completely symmetric with respect to interchanges of its arguments and ψ_{10}^* and ψ_{12}^* denote completely symmetric combinations:

$$\begin{aligned}\psi_{10}^*(s_1,s_2,s_3) &= \psi_{10}(s_1,s_2,s_3) + \psi_{10}(s_3,s_2,s_1) + \psi_{10}(s_1,s_3,s_2)\\ \psi_{12}^*(s_1,s_2,s_3) &= \psi_{12}(s_1,s_2,s_3) + \psi_{12}(s_2,s_1,s_3) + \psi_{12}(s_1,s_3,s_2).\end{aligned} \tag{6.32}$$

The assumption that deformation is unaffected by hydrostatic pressure allows a constitutive relation to be obtained which connects $\mathbf{E}$ with the deviatoric stress $\mathbf{s}$. This is obtained by setting

$$\mathbf{Q} = \mathbf{E}, \qquad \mathbf{P} = \mathbf{R}^T \mathbf{s}\mathbf{R}. \tag{6.33}$$

With this definition $T_\alpha = 0$ and (4.28) reduces to

$$\begin{aligned}\mathbf{E}(t) = \int_{-\infty}^{t} \psi_2 \mathbf{M}_1\, d\tau_{(1)} + \iint_{-\infty}^{t} \{\mathbf{I}\psi_4 T_{12} + \psi_6 \mathbf{M}_1\mathbf{M}_2\}\, d\tau_{(2)}\\ + \iiint_{-\infty}^{t} \{\mathbf{I}\psi_7 T_{123} + \psi_{10} T_{12}\mathbf{M}_3 + \psi_{12}\mathbf{M}_1\mathbf{M}_2\mathbf{M}_3\}\, d\tau_{(3)}\end{aligned} \tag{6.34}$$

where attention is restricted to the third-order theory in $\mathbf{M}$. The condition of zero volume change is applied by forming the quantities $\operatorname{tr}\mathbf{E}$, $\operatorname{tr}\mathbf{E}^2$ and $\operatorname{tr}\mathbf{E}^3$ from (6.34) and substituting in (6.4). This yields an equation which can be written in the form

$$\iint_{-\infty}^{t} \{3\psi_4 + \psi_6 - \psi_2(s_1)\psi_2(s_2)\} T_{12}\, d\tau_{(2)}$$
$$+ \iiint_{-\infty}^{t} \{3\psi_7 + \psi_{12} - 2\psi_2(s_1)\psi_6(s_2, s_3)$$
$$+ \tfrac{4}{3}\psi_2(s_1)\psi_2(s_2)\psi_2(s_3)\} T_{123}\, d\tau_{(3)} = 0 \tag{6.35}$$

where $s_\alpha = t - \tau_\alpha$. Since this equation is supposed to hold for all values of $\mathbf{s}$ it follows that

$$3\psi_4 + \psi_6 = \psi_2(s_1)\psi_2(s_2)$$
$$3\psi_7 + \psi_{12} = \psi_2(s_1)\psi_6(s_2, s_3) + \psi_2(s_3)\psi_6(s_1, s_2) - \tfrac{4}{3}\psi_2(s_1)\psi_2(s_2)\psi_2(s_3) \tag{6.36}$$

where the symmetry properties (4.30) have been invoked. Equations (6.31), (6.34) and (6.36) can now be used to write the constitutive relation in the form

$$\mathbf{E}(t) = \int_{-\infty}^{t} \psi_2 \mathbf{M}_1\, d\tau_{(1)} + \iint_{-\infty}^{t} \{\tfrac{1}{3}\mathbf{I}\psi_2(s_1)\psi_2(s_2) T_{12} + \psi_6(\mathbf{M}_1\mathbf{M}_2 - \tfrac{1}{3}\mathbf{I}\,T_{12})\}\, d\tau_{(2)}$$
$$+ \iiint_{-\infty}^{t} \{2\psi_2(s_1)\psi_6(s_2, s_3) - \tfrac{4}{3}\psi_2(s_1)\psi_2(s_2)\psi_2(s_3)\}\tfrac{1}{3}\mathbf{I}T_{123}\, d\tau_{(3)}$$
$$+ \iiint_{-\infty}^{t} \{\psi_{10} T_{12}\mathbf{M}_3 + \psi_{12}(\mathbf{M}_1\mathbf{M}_2\mathbf{M}_3 - \tfrac{1}{3}\mathbf{I}T_{123})\}\, d\tau_{(3)} \tag{6.37}$$

where ψ_{12} is given by

$$9\psi_{12} = 7\psi_2(s_1)\psi_6(s_2, s_3) + 7\psi_2(s_3)\psi_6(s_1, s_2) - 2\psi_2(s_2)\psi_6(s_1, s_3)$$
$$- 8\psi_2(s_1)\psi_2(s_2)\psi_2(s_3) + 3\psi_{10}^{*}. \tag{6.38}$$

In (6.38) ψ_{10}^{*} is defined by (6.32) and ψ_{12} has the symmetry demanded by (4.30). Because of (6.38), the constitutive relation (6.37) contains only three independent material functions ψ_2, ψ_6 and ψ_{10}, that is, one of each order. This is one less than the number of stress-relaxation functions required in the formulation given in Section 2 and there is a consequent simplification in the experimental programme required for the creep formulation. It is not too surprising that the two formulations lead to different numbers of material functions, since third-order relations in the two formulations are not of equal accuracy and, in addition, the assumption of incompressibility is not a dual restriction as far as $\boldsymbol{\sigma}$ and $\mathbf{E}$ are concerned.

B. Experimental Programme for Determining Creep Functions

In this case all of the material functions can be determined in experiments for which $\mathbf{R} = \mathbf{I}$; this is a great practical advantage. The complete set of tests is shown in Table 4 in which the stress input is imposed on the material and all other stress components are zero. Since the material is incompressible it is not necessary to measure all of the direct strain components; some can be computed using (6.2).

Table 4. Creep experiments for an incompressible material (after Lockett and Stafford, 1969)

Input	Number of experiments	Outputs to be measured
1 $\sigma_{11} = A_1 H(t)$	1	E_{11}
2 $\sigma_{11} = A_2 H(t)$	1	
3 $\sigma_{11} = A_1 H(t) + BH(t-k)$	N	
4 $\sigma_{11} = A_2 H(t) + BH(t-k)$	N	
5 $\sigma_{11} = AH(t),\ \sigma_{22} = BH(t-k)$	N	E_{11}, E_{22}
6 $\sigma_{11} = AH(t),\ \sigma_{22} = BH(t-k) + CH(t-l)$	$\frac{1}{2}N(N-1)$	
7 $\sigma_{11} = AH(t-k),\ \sigma_{22} = BH(t) + CH(t-l)$	$\frac{1}{2}N(N-1)$	

Unlike the experimental programmes required for the formulations discussed in Sections 1 and 2, the present programme appears to be acceptable from an experimental standpoint. All of the tests may be performed on a biaxial testing rig similar to those described by Rivlin and Saunders (1951) and McGuirt and Lianis (1967b) where a sheet of material is subjected to biaxial loading in its own plane. The sheet occupies the (X_1, X_2)-plane and loads are applied in the X_1- and X_2-directions. Facilities must be provided for measuring the in-plane strains E_{11} and E_{22} (for example, by observing a grid inscribed on the sheet) and the other direct strain E_{33}, which would be difficult to measure accurately, can be calculated from (6.2).

The total number of experiments required in the present programme is $N^2 + 2N + 2$; this is less than twice the number required to describe one-dimensional behaviour (see Chapter 5, Section 2) and is approximately a quarter of the number required for a compressible material (see Section 1).

Thus, even though the experimental task is still considerable for an adequate value of $N = 10$ say, the programme is feasible, and has the advantage of requiring only a single testing rig used in a conventional manner.

C. Analysis of Experimental Data

In tests 1 and 2 of Table 4 only E_{11} need be measured, since $E_{22} = E_{33}$ and (6.2) gives

$$2E_{33} = 2E_{22} = (1 + 2E_{11})^{-1/2} - 1 \tag{6.39}$$

from which $\operatorname{tr} \mathbf{E}$ is readily calculated. Since σ_{11} is the only non-zero stress component $s_{11} = \frac{2}{3}\sigma_{11}$, $s_{22} = s_{33} = -\frac{1}{3}\sigma_{11}$ and (6.37) gives

$$27 \operatorname{tr} \mathbf{E} = 18A^2[\psi_2(t)]^2 + 12A^3[\psi_2(t)\,\psi_6(t,t) - \tfrac{2}{3}\{\psi_2(t)\}^3] \tag{6.40}$$

so that the results of tests 1 and 2 determine $\psi_2(t)$ and $\psi_6(t,t)$. In addition when use is made of (6.37) and (6.38) the expression for E_{11} gives an equation from which $\psi_{10}(t,t,t)$ may be found.

Experiment 3 and the expression for $\operatorname{tr} \mathbf{E}$ now yield an equation from which $\psi_6(t, t-k)$ can be calculated and, when repeated for a number of values of k, this determines ψ_6 completely. At this stage only ψ_{10} is not known completely, and (6.37) and (6.38) can be used to write

$$\mathbf{E}(t) = \mathbf{G}(t) + \iiint_{-\infty}^{t} \{\psi_{10}\,T_{12}\,\mathbf{M}_3 + \tfrac{1}{3}\psi_{10}^{*}(\mathbf{M}_1\,\mathbf{M}_2\,\mathbf{M}_3 - \tfrac{1}{3}\mathbf{I}T_{123})\}\,d\tau_{(3)} \tag{6.41}$$

where $\mathbf{G}(t)$ is a functional in which the creep functions are known. It is now readily verified that experiments 3 and 4 of Table 4 give equations for E_{11} which are sufficient to determine ψ_{10}^{*} when two of its arguments are equal. The values of ψ_{10} when two of its arguments are equal are now obtained from measurements of E_{11} and E_{22} in experiment 5, and finally its values for unequal arguments are derived from similar measurements in experiments 6 and 7.

4. INCOMPRESSIBLE PLANE STRAIN

A. Derivation of Constitutive Relation

In this and the following section, simplifications which result from restricting attention to conditions of plane strain and plane stress are considered; the analysis is due to Lockett and Stafford (1969). It is convenient to define $\mathbf{P}$ to be the matrix (stress or strain) for which plane conditions prevail. Thus the formulation for incompressible plane strain is a stress-relaxation formulation obtained as a special case of (6.6).

In order to eliminate the pressure p from the left-hand side of (6.6) an expression is formed for $\mathbf{Q}-\frac{1}{3}\mathbf{I}\,\mathrm{tr}\,\mathbf{Q}$ which yields

$$\mathbf{R}^T\mathbf{sR} = \int_{-\infty}^{t} \psi_2(\mathbf{M}_1-\tfrac{1}{3}\mathbf{I}T_1)\,d\tau_{(1)} + \iint_{-\infty}^{t} \psi_6(\mathbf{M}_1\mathbf{M}_2-\tfrac{1}{3}\mathbf{I}T_{12})\,d\tau_{(2)}$$

$$+\iiint_{-\infty}^{t} [\psi_{10}T_{12}\mathbf{M}_3+\psi_{12}(\mathbf{M}_1\mathbf{M}_2\mathbf{M}_3-\tfrac{1}{3}\mathbf{I}T_{123})]\,d\tau_{(3)} \qquad (6.42)$$

where use has been made of (6.4) in order to eliminate a term in ψ_{10}. For conditions of plane strain coordinate axes are chosen so that

$$\bar{\mathbf{M}} = \begin{Vmatrix} M_{11} & M_{12} \\ M_{21} & M_{22} \end{Vmatrix}, \qquad M_{\alpha 3}=M_{3\alpha}=0 \qquad (\alpha=1,2,3) \qquad (6.43)$$

and the bar notation will be used to denote the corresponding minors of other matrices. When (6.43) is substituted into (6.42) there results a pair of equations for $\bar{\mathbf{R}}^T\bar{\mathbf{s}}\bar{\mathbf{R}}$ and s_{33} (since $R_{\beta 3}=R_{3\beta}=0$, $\beta=1,2$; $R_{33}=1$) in terms of $\bar{\mathbf{M}}$:

$$\bar{\mathbf{R}}^T\bar{\mathbf{s}}\bar{\mathbf{R}} = \int_{-\infty}^{t} \psi_2(\bar{\mathbf{M}}_1-\tfrac{1}{3}\bar{\mathbf{I}}T_1)\,d\tau_{(1)} + \iint_{-\infty}^{t} \psi_6(\bar{\mathbf{M}}_1\bar{\mathbf{M}}_2-\tfrac{1}{3}\bar{\mathbf{I}}T_{12})\,d\tau_{(2)}$$

$$+\iiint_{-\infty}^{t} [\psi_{10}\bar{T}_{12}\bar{\mathbf{M}}_3+\psi_{12}(\bar{\mathbf{M}}_1\bar{\mathbf{M}}_2\bar{\mathbf{M}}_3-\tfrac{1}{3}\bar{\mathbf{I}}\bar{T}_{123})]\,d\tau_{(3)} \qquad (6.44)$$

$$-3s_{33} = \int_{-\infty}^{t} \psi_2\bar{T}_1\,d\tau_{(1)} + \iint_{-\infty}^{t} \psi_6\bar{T}_{12}\,d\tau_{(2)} + \iiint_{-\infty}^{t} \psi_{12}\bar{T}_{123}\,d\tau_{(3)}. \qquad (6.45)$$

Since $\bar{\mathbf{M}}$ is a 2×2 matrix some simplification of these equations can be made through use of the following results for 2×2 matrices (Rivlin, 1955):

$$\mathbf{AB}+\mathbf{BA}-\mathbf{A}\,\mathrm{tr}\,\mathbf{B}-\mathbf{B}\,\mathrm{tr}\,\mathbf{A}+(\mathrm{tr}\,\mathbf{A}\,\mathrm{tr}\,\mathbf{B}-\mathrm{tr}\,\mathbf{AB})\mathbf{I}=0 \qquad (6.46)$$

$$\mathrm{tr}\,\mathbf{ABC}+\mathrm{tr}\,\mathbf{CBA}-\mathrm{tr}\,\mathbf{A}\,\mathrm{tr}\,\mathbf{BC}-\mathrm{tr}\,\mathbf{B}\,\mathrm{tr}\,\mathbf{CA}-\mathrm{tr}\,\mathbf{C}\,\mathrm{tr}\,\mathbf{AB}+\mathrm{tr}\,\mathbf{A}\,\mathrm{tr}\,\mathbf{B}\,\mathrm{tr}\,\mathbf{C}=0 \qquad (6.47)$$

$$\mathbf{ABC}+\mathbf{CBA}=\mathbf{A}\,\mathrm{tr}\,\mathbf{BC}+\mathbf{C}\,\mathrm{tr}\,\mathbf{AB}+(\mathbf{B}-\mathbf{I}\,\mathrm{tr}\,\mathbf{B})(\mathrm{tr}\,\mathbf{A}\,\mathrm{tr}\,\mathbf{C}-\mathrm{tr}\,\mathbf{AC}). \qquad (6.48)$$

When it is noted that $\psi_6=S(1,2)$ equation (6.46) may be used to write

$$\iint_{-\infty}^{t} \psi_6\bar{\mathbf{M}}_1\bar{\mathbf{M}}_2\,d\tau_{(2)} = \iint_{-\infty}^{t} \{\psi_6\bar{T}_1\bar{\mathbf{M}}_2-\tfrac{1}{2}\bar{\mathbf{I}}\psi_6\bar{T}_1\bar{T}_2+\tfrac{1}{2}\bar{\mathbf{I}}\psi_6\bar{T}_{12}\}\,d\tau_{(2)}. \qquad (6.49)$$

Now $\operatorname{tr}\mathbf{E}^{\alpha} = \operatorname{tr}\bar{\mathbf{E}}^{\alpha}$ so that (6.4) and (6.47) yield, to third order,

$$\operatorname{tr}\bar{\mathbf{E}} = \operatorname{tr}\bar{\mathbf{E}}^2, \qquad \operatorname{tr}\bar{\mathbf{E}}^3 = 0 \tag{6.50}$$

and these results allow the right side of (6.49) to be written in the form

$$\tfrac{1}{2}\bar{\mathbf{I}} \iint_{-\infty}^{t} \psi_6 \bar{T}_{12}\, d\tau_{(2)} + \iiint_{-\infty}^{t} \{\psi_6(s_1, s_3)\, H(s_2 - s_1) + \psi_6(s_2, s_3)\, H(s_1 - s_2)\}\, \bar{T}_{12} \bar{\mathbf{M}}_3\, d\tau_{(3)} \tag{6.51}$$

In a similar manner other terms in (6.44) and (6.45) may be re-expressed and these equations finally become

$$\bar{\mathbf{R}}^T \bar{\mathbf{s}} \bar{\mathbf{R}} = \int_{-\infty}^{t} \psi_2 \bar{\mathbf{M}}_1\, d\tau_{(1)} + \bar{\mathbf{I}} \iint_{-\infty}^{t} \phi_6 \bar{T}_{12}\, d\tau_{(2)} + \iiint_{-\infty}^{t} \phi_{10} \bar{T}_{12} \bar{\mathbf{M}}_3\, d\tau_{(3)} \tag{6.52}$$

$$s_{33} = -\int_{-\infty}^{t} \psi_2 \bar{T}_1\, d\tau_{(1)} - 2 \iint_{-\infty}^{t} \phi_6 \bar{T}_{12}\, d\tau_{(2)} \tag{6.53}$$

to third order, where

$$3\phi_6 = \tfrac{1}{2}\psi_6 - \psi_2(s_1)\, H(s_2 - s_1) - \psi_2(s_2)\, H(s_1 - s_2) \tag{6.54}$$

$$\phi_{10} = \psi_{10} + \tfrac{1}{2}\{\psi_{12} + \psi_{12}(s_2, s_1, s_3) - \psi_{12}(s_1, s_3, s_2)\} + \psi_6(s_1, s_3)\, H(s_2 - s_1) + \psi_6(s_2, s_3)\, H(s_1 - s_2) \tag{6.55}$$

from which it follows that $\phi_6 = S(1,2)$ and $\phi_{10} = S(1,2)$.

The constitutive relations (6.52) and (6.53) for incompressible plane strain contain three material functions ψ_2, ϕ_6 and ϕ_{10} which is one fewer than the corresponding case when the assumption of plane strain is not made. The experimental task associated with (6.52) and (6.53) is therefore simpler than that associated with (6.6).

B. Experimental Programme for Incompressible Plane Strain

All of the material functions appearing in (6.52) and (6.53) may be determined in experiments for which $\mathbf{R} = \mathbf{I}$; this is a considerable experimental advantage. The complete set of tests is shown in Table 5; the strain inputs are imposed on the material and E_{22} is allowed to take the value required by (6.2), all other strains are held at zero. Since E_{11} and E_{22} are the only non-zero strains the choice of axes is consistent with the assumption (6.43). In each of the experiments it is necessary to measure σ_{11} and σ_{33}; the experiments are designed so that $\sigma_{22} \equiv 0$.

The tests indicated in Table 5 can be carried out on a conventional rig for performing biaxial tests on a sheet of material. Axes are chosen so that the sheet lies parallel to the (x_1, x_3)-plane and E_{33} is held at zero whilst the required strain input E_{11} is applied. The stresses σ_{11} and σ_{33} required to maintain these strains must be measured during the experiment. There is no need to *apply* the strain E_{22} as it automatically takes the value required by (6.2) provided that the material is incompressible.

The total number of experiments is $N^2 + N + 1$ and this is of the same order of magnitude as the programme for the creep formulation for incompressible materials (see Section 3 where, however, the restriction to plane strain was not made). In fact both programmes can be performed on the same biaxial testing rig provided that it is designed for creep and stress-relaxation experiments.

Table 5. Programme of stress-relaxation tests (after Lockett and Stafford, 1969)

Input	Number of experiments	Outputs to be measured
1 $E_{11} = AH(t)$	1	σ_{11}, σ_{33}
2 $E_{11} = A_1 H(t) + BH(t-k)$	N	
3 $E_{11} = A_2 H(t) + BH(t-k)$	N	
4 $E_{11} = A_1 H(t) + BH(t-k) + CH(t-l)$	$\frac{1}{2}N(N-1)$	
5 $E_{11} = A_2 H(t) + BH(t-k) + CH(t-l)$	$\frac{1}{2}N(N-1)$	

As in Section 3, the present discussion provides the basis for a feasible programme; no experimental investigation using this analysis has been reported at the present time.

C. Analysis of Experimental Data

Since $\sigma_{22} \equiv 0$, all of the non-zero deviatoric stresses s_{11}, s_{22} and s_{33} can be determined from the measured values of σ_{11} and σ_{33}. For experiment 1 of Table 5

$$E_{11} = AH(t), \qquad E_{22} = XH(t), \qquad X = -A/(1+2A) \tag{6.56}$$

and (6.52) and (6.53) yield three equations

$$\begin{aligned} s_{11} &= A\psi_2(t) + (A^2 + X^2)\phi_6(t,t) + A(A^2 + X^2)\phi_{10}(t,t,t) \\ s_{22} &= X\psi_2(t) + (A^2 + X^2)\phi_6(t,t) + X(A^2 + X^2)\phi_{10}(t,t,t) \\ s_{33} &= -(A^2 + X^2)\,[2\phi_6(t,t) + \psi_2(t)] \end{aligned} \tag{6.57}$$

from which $\psi_2(t)$, $\phi_6(t,t)$ and $\phi_{10}(t,t,t)$ may be determined.

Experiments 2 and 3 require that E_{22} is of the form

$$E_{22} = XH(t) + YH(t-k) \tag{6.58}$$

where X and Y are given by (6.2). Equation (6.53) now yields

$$s_{33} + F_3 = -4(AB + XY)\phi_6(t, t-k) \tag{6.59}$$

where F_3 is a function of previously determined quantities and symmetry of ϕ_6 has been used. Equation (6.59) determines ϕ_6 completely when the experiment is repeated for a range of values of k. In a similar manner (6.52) yields expressions for s_{11} and s_{22} which allow

$$\phi_{10}(t, t-k, t-k), \qquad \phi_{10}(t-k, t-k, t), \qquad \phi_{10}(t, t, t-k),$$
$$\phi_{10}(t-k, t, t) \tag{6.60}$$

to be determined from the results of experiments 2 and 3. After repetition for a number of values of k, ϕ_{10} is determined completely when two of its arguments are equal. Finally experiments 4 and 5 give two pairs of equations for s_{11} and s_{22} from which values of ϕ_{10} can be found for arguments which are all distinct. This completes the determination of the stress-relaxation functions ψ_2, ϕ_6 and ϕ_{10}.

5. INCOMPRESSIBLE PLANE STRESS

In this case the creep formulation is chosen so that $\mathbf{M}$ involves the matrix for which plane conditions prevail. Since $\mathbf{P} = \mathbf{R}^T\mathbf{s}\mathbf{R}$ in (6.37), $M_{33} = \dot{P}_{33} = \dot{s}_{33} \neq 0$ and, noting that $\operatorname{tr}\mathbf{M} = 0$, it follows that

$$M_{33} = -\bar{T} \tag{6.61}$$

where the bar has the same significance as in Section 4. It follows that

$$T_{12} = \bar{T}_{12} + \bar{T}_1\bar{T}_2, \quad T_{123} = \bar{T}_{123} - \bar{T}_1\bar{T}_2\bar{T}_3. \tag{6.62}$$

When (6.61) and (6.62) are substituted into (6.37) there result two equations for $\bar{\mathbf{E}}(t)$ and $E_{33}(t)$ from which a number of redundant terms can be removed using (6.46) to (6.48). However this process does not lead to any reduction in the number of material functions contained in the theory and the new equations have no advantages over (6.37). Thus there is no practical simplification in the creep formulation for an incompressible material when conditions of plane stress are assumed.

Chapter 7 Single-Integral Constitutive Relations

The multiple-integral constitutive relations considered in the two previous chapters involve material functions which depend on several integration variables. Experimental determination of these multi-argument functions requires a large number of tests even when a third-order theory is adequate, and such a direct approach would be prohibitive if higher-order nonlinearity was involved. In order to consider theories which do not require large experimental programmes, it is natural to propose special constitutive relations which involve only a single integration, and to test the ability of these theories to model experimental data. If the nonlinearity is expressed in an explicit form, the material functions depend only on a single time-like variable, provided that the material is of the hereditary type. In principle the number of tests required for the determination of material properties is equal to the number of material functions but, in practice, further tests are required to improve accuracy and determine the range of validity of the theory.

Consideration of a single-integral relation limits the form of nonlinearity modelled by the theory, and it should not be expected that such a theory can be adequate in all situations. Whereas certain single-integral relations have been shown to be capable of representing experimental data, no criteria are currently available for determining which theory is most appropriate to a given situation. For this reason it is important to note the occasions where poor modelling is obtained, although this information is rarely reported in the literature.

Nonlinearity in a single-integral relation may be achieved in a number of ways, not all of which are considered in the present chapter. Here a discussion is given of the theories proposed by Bernstein, Kearsley and Zapas (BKZ) and by Lianis and his co-workers. A considerable amount of experimental data has been analysed in the context of these theories and they have been shown to be adequate in a significant number of situations. Although they contain some elements of empiricism, they are based on rational arguments, and they have been formulated to model three-dimensional behaviour. Other single-integral theories are considered in Chapter 9, but these are derived on a semi-empirical basis, and generally have been considered only in one-dimensional form.

1. BKZ THEORIES

Two forms of single-integral constitutive relation are associated with Bernstein, Kearsley and Zapas (see Bernstein *et al.*, 1963). They noted that if a material is observed at sufficiently low temperatures and for short times, it is difficult to distinguish its behaviour from that of an elastic solid. In these circumstances it is appropriate to choose a constitutive relation which, like that for an elastic solid, is defined with respect to a preferred configuration. When the material is observed for sufficiently long times and at sufficiently high temperatures, flow behaviour is more significant and a fluid-like theory may be appropriate. The BKZ solid and fluid theories are described in the next two sub-sections and the analysis of experimental data is considered in Section 1C.

A. BKZ Theory for an Incompressible Solid

This theory is readily derived from the general form (4.8) when the argument of the functional is changed to the strain measure $\mathbf{E}(\tau)$. If the condition of incompressibility is supposed to limit definition of stress to within a hydrostatic pressure (see discussion on pages 99–100), the constitutive relation has the form

$$\boldsymbol{\sigma}(t) = -p\mathbf{I} + \mathbf{F}(t) \underset{\tau=-\infty}{\overset{t}{\mathscr{F}}} \{\mathbf{E}(\tau)\} \mathbf{F}^T(t) \tag{7.1}$$

where $\mathscr{F}$ is a matrix functional. For an initially isotropic material an integral polynomial approximation for $\mathscr{F}$ which is linear in $\mathbf{E}$ is

$$\underset{\tau=-\infty}{\overset{t}{\mathscr{F}}} \{\mathbf{E}(\tau)\} = M\mathbf{I} + \mathbf{I} \int_{-\infty}^{t} \bar{A}(t-\tau) \operatorname{tr} \mathbf{E}(\tau)\, d\tau + 2 \int_{-\infty}^{t} \bar{B}(t-\tau)\, \mathbf{E}(\tau)\, d\tau \tag{7.2}$$

where M is a constant and $\bar{A}$ and $\bar{B}$ are functions of their indicated arguments. They are not stress-relaxation functions since, in order to align with Bernstein *et al.* (1963), $\mathbf{E}$ has been written rather than $\dot{\mathbf{E}}$ in the integrands. It is not necessary to write explicitly terms involving $\operatorname{tr}\mathbf{E}(t)$ and $\mathbf{E}(t)$ in (7.2) provided that $\bar{A}$ and $\bar{B}$ are allowed to contain Dirac delta functions.

In a single-step stress-relaxation test which is not confined to any particular component of $\mathbf{E}$, the strain matrix has the form

$$\mathbf{E}(t) = \mathbf{E}_0\, H(t) \tag{7.3}$$

where $\mathbf{E}_0$ is independent of t, and (7.2) gives

$$\underset{\tau=-\infty}{\overset{t}{\mathscr{F}}} \{\mathbf{E}(\tau)\} = M\mathbf{I} + \mathbf{I}A(t) \operatorname{tr} \mathbf{E}_0 + 2B(t)\, \mathbf{E}_0 \tag{7.4}$$

where

$$A(t) = \int_{0}^{t} \bar{A}(\tau)\, d\tau, \qquad B(t) = \int_{0}^{t} \bar{B}(\tau)\, d\tau. \tag{7.5}$$

It follows from (7.1) and (7.4) that the relation between the equilibrium stress $\boldsymbol{\sigma} = \boldsymbol{\sigma}(\infty)$ and $\mathbf{E}_0$ can be expressed as

$$\boldsymbol{\sigma} = -p\mathbf{I} + \mathbf{F}\mathbf{W}'\mathbf{F}^T \tag{7.6}$$

where $\mathbf{W}' = \partial W/\partial \mathbf{E}_0$ and

$$W = M \operatorname{tr} \mathbf{E}_0 + \tfrac{1}{2}A(\operatorname{tr} \mathbf{E}_0)^2 + \mathbf{B} \operatorname{tr} \mathbf{E}_0^2 \tag{7.7}$$

where $A = A(\infty)$, $B = B(\infty)$. In terms of the invariants I_1 and I_2 commonly used in the theory of incompressible nonlinear elasticity (see 4.46)

$$I_1 = \operatorname{tr} \mathbf{C}, \qquad I_2 = \tfrac{1}{2}(I_1^2 - \operatorname{tr} \mathbf{C}^2) \tag{7.8}$$

(7.7) can be re-expressed as

$$W = (\tfrac{1}{2}M + B)(I_1 - 3) + \tfrac{1}{4}(\tfrac{1}{2}A + B)(I_1 - 3)^2 - \tfrac{1}{2}B(I_2 - 3). \tag{7.9}$$

This expression indicates that the long-term equilibrium behaviour of the material is identical to that considered by Signiorini (1955); and so (7.1) with (7.2) is referred to as the Signiorini form of the BKZ theory. If $A + 2B = 0$ the long-term behaviour is that of a Mooney material (Mooney, 1940), whereas if $A = B = 0$ it is neo-Hookian (see Chapter 4, Section 4).

It is of interest now to consider the predictions of the constitutive relation (7.1) with (7.2) for the stress in a simple extension stress-relaxation test. A rod or strip of material is loaded so that the only non-zero stress component is $\sigma_{11}(t) = \sigma(t)$ and the form of $\sigma(t)$ is controlled so that the extension in the x_1-direction is zero up to some time $t = 0$ and takes on a constant non-zero value thereafter. The deformation in the material can be expressed in the form

$$x_1 = X_1[1 + (\lambda - 1)\,H(t)], \qquad x_2 = X_2[1 + (\lambda^{-1/2} - 1)\,H(t)]$$
$$x_3 = X_3[1 + (\lambda^{-1/2} - 1)\,H(t)] \tag{7.10}$$

where λ is the extension ratio in the x_1-direction; the extension ratios in the x_2 and x_3-directions are $\lambda^{-1/2}$ due to the assumption of incompressibility. When (7.10) is substituted into (7.1) and (7.2) the pressure p can be determined from the fact that $\sigma_{22} = \sigma_{33} = 0$, and the expression for $\sigma_{11} = \sigma(t)$ can then be written in the form

$$\sigma(t)/(\lambda^2 - \lambda^{-1}) = (\lambda^2 - 1)\,[\tfrac{1}{2}A(t) + B(t)] + \lambda^{-1}[A(t) + B(t)] + M - A(t). \tag{7.11}$$

This equation will be considered again in Sub-section C, where the analysis of experimental data is discussed.

Before proceeding to consider the BKZ fluid theory it is necessary to make one further comment concerning the derivation of the solid-type theory (7.1) with (7.2). Expression (7.2) is a linearized form of $\mathscr{F}$ in which terms of second and higher order in $\mathbf{E}$ have been omitted; there are two alternative ways in which this linearization can be justified. First it may be postulated that the material is such that material functions appearing in higher-order terms may be neglected. Alternatively (7.2) may be considered to be an expression which is a valid approximation for sufficiently small $\mathbf{E}$. However, since the BKZ theory is for an incompressible material, and since for small $\mathbf{E}$ equation (6.4) shows that $\mathrm{tr}\,\mathbf{E}$ is a second-order quantity, it would then have been logical to omit the term in $\mathrm{tr}\,\mathbf{E}$ from (7.2). Bernstein *et al.* (1963) did not do this: in fact their method of analysis of experimental data (see Sub-section C) is dependent upon the term in $\mathrm{tr}\,\mathbf{E}$ being significant. Thus (7.2) must be motivated by the first alternative above, that is, by the desire to consider a constitutive relation involving single integrals only. Justification for such a model can come only from its proven ability to model experimental data.

B. BKZ Theory for an Incompressible Elastic Fluid

In their postulated theory for an elastic fluid Bernstein *et al.* (1963) use a deformation measure $\mathbf{F}_\tau$ which is inverse to (4.18):

$$\mathbf{F}_\tau = \left\| \frac{\partial x_i(t)}{\partial x_j(\tau)} \right\| \tag{7.12}$$

in the sense that the roles of t and τ have been reversed and for which the corresponding strain matrix $\mathbf{E}_\tau$ is

$$\mathbf{E}_\tau = \tfrac{1}{2}(\mathbf{F}_\tau^T \mathbf{F}_\tau - \mathbf{I}). \tag{7.13}$$

They then postulate a constitutive relation of the form

$$\boldsymbol{\sigma}(t) = -p\mathbf{I} + \int_{-\infty}^{t} \mathbf{F}_\tau \mathbf{U}' \mathbf{F}_\tau^T \, d\tau \tag{7.14}$$

where p is a hydrostatic pressure and

$$U'_{ij} = \partial U / \partial E_{\tau ij}, \qquad U = U[\mathbf{E}_\tau(t, \tau), t - \tau]. \tag{7.15}$$

If the integral in (7.14) is disregarded, that equation is equivalent to the constitutive relation for an incompressible elastic material (see Chapter 4, Section 4) in which the configuration at time τ is the preferred configuration. The equation satisfies the requirements of the principle of frame-indifference provided that the elastic potential U depends on $\mathbf{E}_\tau$ only through its invariants. The BKZ equation (7.14) assumes that the effect of the configuration at time $\tau < t$ on the stress at time t is equivalent to the effect of stored elastic energy with the configuration at time τ as the preferred configuration, the total stress being the integrated sum of all contributions from all $\tau < t$.

When U is expanded in powers of the invariants $\operatorname{tr}\mathbf{E}_\tau$ and $\operatorname{tr}\mathbf{E}_\tau^2$, and terms up to second order in $\mathbf{E}_\tau$ are retained so that U' is linear in $\mathbf{E}_\tau$, there results the expression

$$U = m(t-\tau)\operatorname{tr}\mathbf{E}_\tau + \tfrac{1}{2}a(t-\tau)[\operatorname{tr}\mathbf{E}_\tau]^2 + b(t-\tau)\operatorname{tr}\mathbf{E}_\tau^2. \tag{7.16}$$

where m, a and b are material functions. The constitutive relation (7.14) now reduces to the form

$$\boldsymbol{\sigma}(t) = -p\mathbf{I} + \int_{-\infty}^{t} \{[m(t-\tau) + \tfrac{1}{2}a(t-\tau)(\operatorname{tr}\mathbf{B}_\tau - 3) - b(t-\tau)]\mathbf{B}_\tau + b(t-\tau)\mathbf{B}_\tau^2\}\,d\tau \tag{7.17}$$

where $\mathbf{B}_\tau = \mathbf{F}_\tau\mathbf{F}_\tau^T$ is the left Cauchy-Green strain measure corresponding to (7.12).

In a single-step stress-relaxation situation, in which the material is given an instantaneous deformation at $t=0$ and is then held fixed, the strain measure $\mathbf{B}_\tau$ is equal to the unit matrix $\mathbf{I}$ for $t<0$ and for t, $\tau>0$. When $t>0$ and $\tau<0$, $\mathbf{B}_\tau$ has some constant value $\mathbf{B}_\tau^{(0)}$ independent of both t and τ. Thus, for $t>0$ (7.17) gives

$$\boldsymbol{\sigma}(t) = -p\mathbf{I} + [M(t) + \tfrac{1}{2}A(t)(\operatorname{tr}\mathbf{B}_\tau^{(0)} - 3) - B(t)]\mathbf{B}_\tau^{(0)} + B(t)\mathbf{B}_\tau^{(0)2} \tag{7.18}$$

where

$$M(t) = \int_t^\infty m(\tau)\,d\tau, \qquad A(t) = \int_t^\infty a(\tau)\,d\tau, \qquad B(t) = \int_t^\infty b(\tau)\,d\tau. \tag{7.19}$$

When $t>0$ and $\tau<0$, the simple extension stress-relaxation test considered in (7.10) can be characterized in the present notation by replacing x_i by $x_i(t)$ and X_i by $x_i(\tau)$ in (7.10). It then follows that

$$\mathbf{B}_\tau^{(0)} = \operatorname{diag}\{\lambda^2, \lambda^{-1}, \lambda^{-1}\} \tag{7.20}$$

and (7.18) then gives $\sigma_{11}(t) = \sigma(t)$ in the form

$$\sigma(t)/(\lambda^2 - \lambda^{-1}) = (\lambda^2 - 1)[\tfrac{1}{2}A(t) + B(t)] + \lambda^{-1}[A(t) + B(t)] + M(t) - A(t) \tag{7.21}$$

where use has been made of the fact that $\sigma_{22} = \sigma_{33} = 0$.

Equations (7.21) and (7.11) are similar in form, however, it should be noted that $M(t)$, $A(t)$ and $B(t)$ are not the same in the two expressions. The functions (7.19) necessarily tend to zero at large times whereas there is no such requirement on (7.5); in addition $M(t)$ depends on t in (7.21) but is a constant in (7.11). The same notation has been used in the two equations in order that they may be considered simultaneously in the analysis of experimental data (see Sub-section C).

At large times (7.19) and (7.21) show that $\sigma(t)$ tends to zero, thus there is no long-term equilibrium. In the time scale of an experiment it may not be possible to determine the theoretical long-term behaviour of the material. If, in that time scale, the stress has relaxed to zero, then a fluid-type of theory is appropriate. However, if stresses are non-zero at the longest time of the experiment, then it is not possible to say definitely whether a non-zero long-term equilibrium exists. Thus, although the theory presented in this sub-section is a theory for fluids, it can predict material responses which are usually associated with solids.

In keeping with the remarks made at the end of the previous sub-section it is noted that for small strains in an incompressible material the second term on the right-hand side of (7.16) is of fourth order in $\mathbf{E}_\tau$. Thus the remarks made about the motivation of the solid theory also apply to the motivation of (7.17).

C. Experimental Determination of Material Functions in BKZ Theories

For stress-relaxation in simple extension the stress $\sigma(t)$ required to maintain an extension ratio λ is given by (7.11) or (7.21), both of which can be written in the form

$$\sigma_R(\lambda, t) = (\lambda^2 - 1)\,\alpha(t) + \lambda^{-1}\,\beta(t) + \gamma(t) \tag{7.22}$$

where the reduced stress σ_R and material functions α, β and γ are given by

$$\sigma_R = \sigma/(\lambda^2 - \lambda^{-1}), \qquad \alpha(t) = \tfrac{1}{2}A(t) + B(t)$$

$$\beta(t) = A(t) + B(t), \qquad \gamma(t) = M(t) - A(t). \tag{7.23}$$

If the stress-relaxation experiment is repeated for a range of values of λ then it is possible to plot a set of isochronous curves of $\sigma_R(t)$ against λ^{-1} for various values of t; a typical curve is shown schematically in Fig. 1. It is readily shown using (7.22) that these curves have stationary values when

$$\lambda^{-1} = [2\alpha(t)/\beta(t)]^{1/3} \tag{7.24}$$

so that the function $R(t) = \alpha(t)/\beta(t)$ can be determined from the positions of the stationary values on the experimental curves: in practice these values are minima which correspond to positive values for $\alpha(t)$.

When $R(t)$ is known it is convenient to write (7.22) in the form

$$\sigma_R(t) = [(\lambda^2 - 1)\,R(t) + \lambda^{-1}]\,\beta(t) + \gamma(t) \tag{7.25}$$

so that isochronous plots of $\sigma_R(t)$ against $[(\lambda^2-1)R(t)+\lambda^{-1}]$ are straight lines with slopes $\beta(t)$ and intercepts $\gamma(t)$ provided, of course, that the experimental data are adequately modelled by (7.22).

Bernstein *et al.* (1963) presented experimental data for three materials: plasticized polyvinyl chloride, polyisobutylene and vulcanized butyl rubber, and they analysed these data according to (7.24) and (7.25). For the PVC data the isochrones of σ_R against λ^{-1} for times $t=2$, 10, 100 min had minima corresponding to extension ratios in the range 1.1 to 1.2. The corresponding isochrones of σ_R against $[(\lambda^2-1)R+\lambda^{-1}]$ gave straight lines within the limits of experimental accuracy, showing that the BKZ theories were capable of modelling PVC in the situations considered. The calculated values of $M(t)$

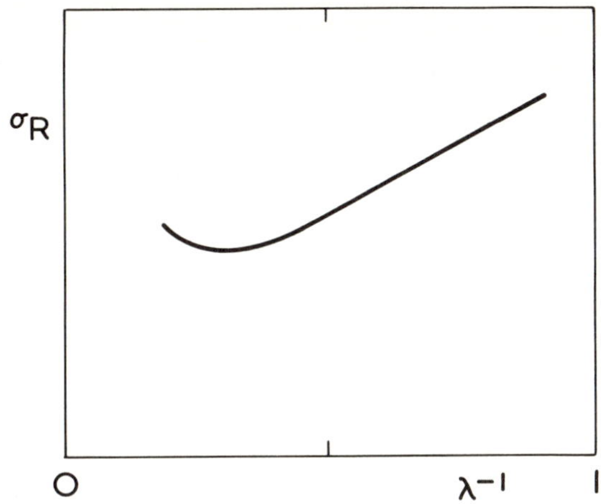

Fig. 1. Schematic diagram of typical isochrone of σ_R against λ^{-1}.

were found to have a small variation with t, suggesting that the fluid theory is more appropriate.

The isochrones of σ_R versus λ^{-1} for polyisobutylene (at −50°C and 25°C) showed no minima in the range of extension ratios considered. Bernstein *et al.* (1963) concluded that the minima would appear at higher extension ratios (greater than $\lambda=2$) and therefore chose $R(t)\equiv 0$ to model their data. With this value the isochrones of σ_R against $[(\lambda^2-1)R+\lambda^{-1}]$ again gave good straight lines. Similar conclusions held for data for the vulcanized butyl rubber, although in this case long-term results (up to 36 hr) were obtained which showed a minimum in the isochrone of σ_R versus λ^{-1} at approximately $\lambda=3.2$. For both materials the function $M(t)$ showed considerable time-dependence, indicating that the fluid theory is more appropriate.

Before concluding the discussion of the experimental results of Bernstein *et al.* (1963) it is interesting to make some comments on the size of the strain measure **E**. For simple extension

$$\operatorname{tr}\mathbf{E} = \tfrac{1}{2}\lambda^2 + \lambda^{-1} - \tfrac{3}{2}, \qquad E_{11}^2 = (\mathbf{E}^2)_{11} = \tfrac{1}{4}(\lambda^2 - 1)^2 \tag{7.26}$$

and the variation with λ of these quantities is shown in Table 1. It is clear that for $\lambda = 3$, $\operatorname{tr}\mathbf{E}$ is order $\mathbf{E}$, whereas $\operatorname{tr}\mathbf{E}$ is order $\mathbf{E}^2$ when $\lambda \leqslant 2$. Since the minima in the isochrones σ_R versus λ^{-1} for polyisobutylene and vulcanized butyl rubber occurred at $\lambda \simeq 3$, it is apparent that it is valid to use the full linearization (7.2) for $\mathscr{F}$. However, for the data on plasticized PVC, the minima occurred at $\lambda \simeq 1.2$ and this value lies in the range where $\operatorname{tr}\mathbf{E}$ is order $\mathbf{E}^2$. Since the term $\operatorname{tr}\mathbf{E}$ is essential to the analysis of Bernstein *et al.* (1963), and since their experimental data is adequately modelled by their theories, this suggests that (7.2) is an adequate *second-order* approximation in the range of experimental data.

Table 1. Variation of $\operatorname{tr}\mathbf{E}$ and E_{11}^2 with λ

λ	1	1.1	1.2	1.3	1.4	1.5	2.0	3.0
$\operatorname{tr}\mathbf{E}$	0	0.02	0.05	0.08	0.19	0.30	1.00	3.3
E_{11}^2	0	0.01	0.05	0.12	0.23	0.40	2.25	16.0

Further experimental support for the BKZ fluid theory was obtained by Zapas and Craft (1965) who obtained data in simple extension for other forms of time-dependent strains. For an arbitrary deformation history

$$x_1(t) = \lambda(t)\, X_1, \qquad x_2(t) = \lambda^{-1/2}(t)\, X_2, \qquad x_3(t) = \lambda^{-1/2}(t)\, X_3 \tag{7.27}$$

defined relative to an arbitrary reference configuration, it is readily shown that

$$\mathbf{B}_\tau = \operatorname{diag}\left\{\frac{\lambda^2(t)}{\lambda^2(\tau)}, \frac{\lambda(\tau)}{\lambda(t)}, \frac{\lambda(\tau)}{\lambda(t)}\right\} \tag{7.28}$$

and for $\sigma_{22} = \sigma_{33} = 0$, (7.17) gives $\sigma_{11} = \sigma(t)$ in the form

$$\sigma(t) = \int_{-\infty}^{t} \left[\left\{m + \tfrac{1}{2}a\left[\frac{\lambda^2(t)}{\lambda^2(\tau)} + 2\frac{\lambda(\tau)}{\lambda(t)} - 3\right] - b\right\}\left\{\frac{\lambda^2(t)}{\lambda^2(\tau)} - \frac{\lambda(\tau)}{\lambda(t)}\right\} + b\left\{\frac{\lambda^4(t)}{\lambda^4(\tau)} - \frac{\lambda^2(\tau)}{\lambda^2(t)}\right\}\right] d\tau \tag{7.29}$$

where m, a and b are evaluated at $t - \tau$.

Using (7.29) and functions $\alpha(t)$, $\beta(t)$ and $\gamma(t)$ calculated from simple stress-relaxation data Zapas and Craft (1965) obtained predictions for material behaviour in double-step and triple-step stress-relaxation tests, in creep and recovery tests, and in constant strain-rate tests. Thus, for example, in a double-step stress-relaxation test characterized by

$$\lambda(t) = \begin{cases} 1 & t < 0 \\ \lambda & 0 \leqslant t < t_1 \\ \eta & t_1 \leqslant t \end{cases} \tag{7.30}$$

(7.29) gives $\sigma(t)$ for times $t > t_1$ in the form

$$\sigma(t) = (\eta^2 - \eta^{-1})\,\sigma_R(\eta, t) + \left(\frac{\eta^2}{\lambda^2} - \frac{\lambda}{\eta}\right)\left\{\sigma_R\left(\frac{\eta}{\lambda}, t - t_1\right) - \sigma_R\left(\frac{\eta}{\lambda}, t\right)\right\} \tag{7.31}$$

where $\sigma_R(\lambda, t)$ is defined by (7.22). Zapas and Craft (1965) showed that the predictions obtained from (7.31), and similar expressions for the other tests, agreed well with their experimental data for a plasticized PVC and for poly-isobutylene. In the prediction for the recovery phase of the creep and recovery test it is necessary to evaluate the function $\sigma_R(\lambda, t)$ at a value $\lambda < 1$, since the material suffers a relative compression when the load is removed. In their calculations Zapas and Craft (1965) assumed that the expression (7.22) evaluated from tests in which $\lambda > 1$ could be used in calculations for which $\lambda < 1$. The predicted and measured recovery data appear to be in good agreement, but the initial stages of recovery, which are difficult to model, are obscured by the logarithmic presentation of data.

For the analysis of creep data Zapas and Craft (1965) used the measured extension ratios to calculate the applied stress which should, of course, be constant. The applied and calculated values differed by up to 4% over the first 6 min after loading. Thereafter the difference between the two values tended to zero. Similar accuracy was obtained in the calculated values of stress or extension ratio in the other tests. Thus the work of Zapas and Craft (1965) gives further support for the BKZ fluid theory although, since their tests were all in simple extension, further comparison between theory and experiment is required to justify its use in more general situations.

2. FINITE LINEAR VISCOELASTICITY

A. The Constitutive Relation: Lianis's Theory

The starting point for the discussion of the present section is the form of constitutive relation (4.20), namely

$$\boldsymbol{\sigma}(t) = \mathop{\mathbf{G}}_{\tau=-\infty}^{t} \{\mathbf{C}_t(\tau); \mathbf{B}(t)\} \tag{7.32}$$

where $\mathbf{G}$ is a matrix functional. This theory has explicit dependence on the current value of the strain $\mathbf{B}(t)$ which is measured with respect to the initial configuration. Dependence on the history of deformation is expressed through a relative strain measure so that, from this point of view, it is the current configuration which is a preferred state rather than the initial reference configuration. This form is particularly suitable when stress has a fairly general dependence on the *current strain* with respect to the initial configuration, but depends on only the recent form of the deformation *history*. A material behaving in this manner would, for example, have the same response to each of the deformation histories shown schematically in Fig. 2 provided that the memory of the material is significant over a time interval less than T. This discussion gives a simple insight into the study of materials with fading memory; these are materials for which current values of stress are most strongly influenced by the recent portions of the deformation history.

Starting from the equation (7.32) Coleman and Noll (1961) derived a constitutive relation for a compressible isotropic material with fading memory in which only linear terms in the strain history are retained. This theory, which they called finite linear viscoelasticity, can be expressed in the form

$$\begin{aligned}\boldsymbol{\sigma}(t) &= -p\mathbf{I} + (\phi + I_1\psi)\mathbf{B} - \psi\mathbf{B}^2 \\ &+ \sum_{\alpha=0}^{2}\int_{-\infty}^{t} \phi_\alpha(t-\tau)\,[\mathbf{B}^\alpha\,\dot{\mathbf{C}}_t(\tau) + \dot{\mathbf{C}}_t(\tau)\,\mathbf{B}^\alpha]\,d\tau \\ &+ \sum_{\alpha=0}^{2}\sum_{\beta=0}^{2}\int_{-\infty}^{t} \phi_{\alpha\beta}(t-\tau)\,\mathbf{B}^\alpha\,\mathrm{tr}\,[\mathbf{B}^\beta\,\dot{\mathbf{C}}_t(\tau)]\,d\tau \qquad (7.33)\end{aligned}$$

where the twelve relaxation functions ϕ_α and $\phi_{\alpha\beta}$ are functions of time and also of the current values of the strain invariants I_1, I_2 and I_3 defined by (4.46). In general p, ϕ and ψ are all material functions depending on the current values of the invariants. The derivation of (7.33) depends on the assumption that the motion of the material in the recent past is slow in some sense. However, it does not follow that these are necessarily the only conditions under which the equation is valid, and it is not obvious from purely theoretical considerations whether a given motion satisfies this condition for a particular material. It will be seen later that (7.33) has provided adequate models of experimental data in conditions of fairly wide generality and at strain rates of the order of 12% sec^{-1} (see Section 2E), so that the requirement of slow motion is not too restrictive.

Lianis (1967) has shown that thermodynamic considerations require the functions $\phi_{\alpha\beta}$ to satisfy the symmetry relation $\phi_{\alpha\beta} = \phi_{\beta\alpha}$, so that only nine of the twelve relaxation functions are independent. For an incompressible material p becomes an undefined hydrostatic pressure and the terms involving ϕ_{00}, ϕ_{01} and ϕ_{02} can be absorbed into this term. In addition, since $I_3 \equiv 1$, the material parameters do not depend on this invariant.

A simplified form of the constitutive relation of finite linear viscoelasticity was proposed by Lianis (1963) and, although this form has been superseded by the modified form discussed in Section 3, it has been used successfully to model a wide range of experimental data. A first assumption in the Lianis (1963) theory is that the dependence of the stress-relaxation functions on the invariants may be neglected; a theory which does not include this restriction has been proposed by Valanis and Landel (1967). The other assumption made in formulating the theory is that the long- and short-term behaviour should be

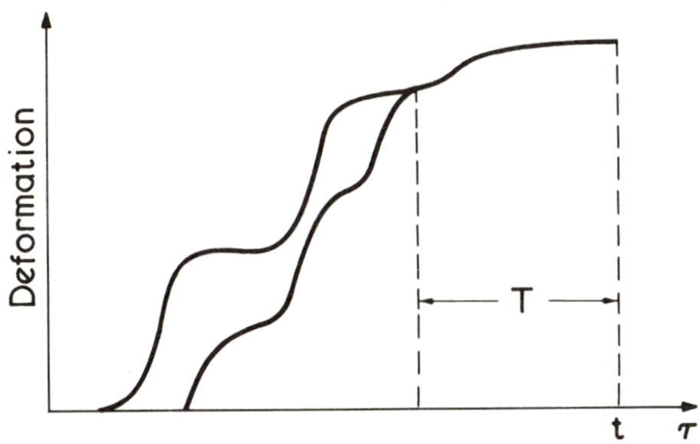

Fig. 2. Two deformation programmes producing same stress $\sigma(t)$ in a material with memory time less than T.

of the simple form modelled by a Mooney elastic material (see Chapter 4, Section 4). It is convenient to define the stress-relaxation functions so that $\phi_\alpha \to 0$, $\phi_{\alpha\beta} \to 0$ as $t \to \infty$, and the first three terms on the right-hand side of (7.33) then represent the long-time elastic behaviour. In order to achieve behaviour of Mooney type it is necessary to set

$$\phi = a + b(I_1 - 3), \qquad \psi = c \tag{7.34}$$

where a, b and c are constants. In order that short-time behaviour should have the same form it is necessary that

$$\phi_{10}, \phi_{12}, \phi_{20}, \phi_{21}, \phi_{22} \tag{7.35}$$

should be set equal to zero: sufficiency of this condition is demonstrated in the next sub-section. It follows from the assumptions made above that the Lianis

(1963) form of constitutive relation for an incompressible isotropic material is

$$\begin{aligned}\boldsymbol{\sigma}(t) = -p\mathbf{I} + [a + b(I_1 - 3) + cI_1]\,\mathbf{B} - c\mathbf{B}^2 \\ + 2\int_{-\infty}^{t} \phi_0(t-\tau)\,\dot{\mathbf{C}}_t(\tau)\,d\tau + \int_{-\infty}^{t} \phi_1(t-\tau)\,[\mathbf{B}\dot{\mathbf{C}}_t(\tau) + \dot{\mathbf{C}}_t(\tau)\,\mathbf{B}]\,d\tau \\ + \int_{-\infty}^{t} \phi_2(t-\tau)\,[\mathbf{B}^2\,\dot{\mathbf{C}}_t(\tau) + \dot{\mathbf{C}}_t(\tau)\,\mathbf{B}^2]\,d\tau + \mathbf{B}\int_{-\infty}^{t} \phi_3(t-\tau)\,\dot{I}_1(\tau)\,d\tau\end{aligned} \tag{7.36}$$

where $\phi_3 \equiv \phi_{11}$ and $I_1(\tau) = \mathrm{tr}\,[\mathbf{B}\mathbf{C}_t(\tau)]$. The predictions of this theory will now be compared with experimental data for several different forms of deformation history.

B. Single-Step Stress-Relaxation

Consider a general single-step stress-relaxation test in which the material is in its preferred state for times $t < 0$ and is given some constant deformation for all times $t > 0$. When $t > 0$

$$x_i(\tau) = X_i, \qquad \tau < 0; \qquad x_i(\tau) = x_i(t), \qquad \tau > 0 \tag{7.37}$$

and it follows that

$$\left.\begin{aligned}\mathbf{C}_t(\tau) &= \mathbf{I}, \qquad \tau > 0 \\ \mathbf{C}_t(\tau) &= \mathbf{B}^{-1} \equiv \mathbf{B}_0^{-1}, \qquad \tau < 0\end{aligned}\right\} \tag{7.38}$$

where $\mathbf{B}_0$ is a constant matrix. Thus

$$\mathbf{C}_t(\tau) = \mathbf{B}_0^{-1} + (\mathbf{I} - \mathbf{B}_0^{-1})\,H(\tau), \qquad \dot{\mathbf{C}}_t(\tau) = (\mathbf{I} - \mathbf{B}_0^{-1})\,\delta(\tau) \tag{7.39}$$

and (7.36) yields

$$\begin{aligned}\boldsymbol{\sigma} = -p\mathbf{I} + [a + b(I_1 - 3) + cI_1 + 2\phi_1(t) - 2\phi_2(t) \\ + (I_1 - 3)\phi_3(t) + 2I_1\,\phi_0(t)]\,\mathbf{B}_0 - [c - 2\phi_2(t) + 2\phi_0(t)]\,\mathbf{B}_0^2\end{aligned} \tag{7.40}$$

where $I_1 = \mathrm{tr}\,\mathbf{B}_0$. In (7.40) terms involving the unit matrix **I** have been absorbed into the pressure term, and the Hamilton-Cayley theorem

$$\mathbf{B}^3 - I_1\,\mathbf{B}^2 + I_2\,\mathbf{B} - I_3\,\mathbf{I} = \mathbf{0} \tag{7.41}$$

has been used in the form

$$\mathbf{B}_0^{-1} = \mathbf{B}_0^2 - I_1\,\mathbf{B}_0 + I_2\,\mathbf{I} \tag{7.42}$$

since $I_3 = 1$.

When $t = 0$ the coefficient of $\mathbf{B}_0$ in (7.40) is linear in I_1, whereas the coefficient of $\mathbf{B}_0^2$ is constant; this is the form required for Mooney behaviour and demonstrates that setting (7.35) equal to zero is sufficient to produce the assumed form of short-time elastic response.

The notation of equation (7.18) is such that $\mathbf{B}_\tau^{(0)} \equiv \mathbf{B}_0$ and it is evident that (7.18) and (7.40) are identical under the set of relations

$$\begin{aligned} A(t) &= 2[b + c + 2\phi_0(t) + \phi_3(t)] \\ B(t) &= -c - 2\phi_0(t) + 2\phi_2(t) \\ M(t) &= a + 2c + 4\phi_0(t) + 2\phi_1(t). \end{aligned} \tag{7.43}$$

Thus, in single-step stress-relaxation situations, the predictions of the BKZ fluid theory and the Lianis theory (7.36) are identical. In particular the Lianis theory predicts the behaviour in simple extension to be governed by (7.22) where

$$\begin{aligned} \alpha(t) &= b + 2\phi_2(t) + \phi_3(t) \\ \beta(t) &= 2b + c + 2\phi_0(t) + 2\phi_2(t) + 2\phi_3(t) \\ \gamma(t) &= a - 2b + 2\phi_1(t) - 2\phi_3(t). \end{aligned} \tag{7.44}$$

Values for the functions α, β and γ for a particular material can be determined using the experimental procedure outlined in Section 1C. Since $\phi_\alpha(\infty) = 0$, it follows that $\alpha(\infty)$, $\beta(\infty)$ and $\gamma(\infty)$ (or, in practice, long-time values) are sufficient to determine a, b and c. For other values of t the experimentally determined values of $\alpha(t)$, $\beta(t)$ and $\gamma(t)$ provide three equations relating $\phi_\alpha(t)$ $(\alpha = 0, 1, 2, 3)$. It follows that one further test is required to determine Lianis's stress-relaxation functions. From the discussion above it is apparent that single-step stress-relaxation cannot provide this extra information; other types of test are discussed in the following sub-sections. First, some additional experimental data are considered which have been modelled by (7.40) or its equivalent form:

$$\boldsymbol{\sigma} = -p\mathbf{I} + [\gamma + 3\beta - 4\alpha + (\beta - \alpha)(I_1 - 3)]\,\mathbf{B}_0 - (\beta - 2\alpha)\,\mathbf{B}_0^2. \tag{7.45}$$

It is recalled that the experimental data of Bernstein *et al.* (1963) for plasticized PVC are modelled adequately by the BKZ theory, and therefore also the Lianis theory, with non-zero values for α, β and γ. For both theories the data for polyisobutylene and butyl rubber are modelled with $\alpha \equiv 0$.

Goldberg and Lianis (1967) considered stress relaxation in combined torsion-tension of a cylindrical rod of an SBR polymer and compared their experimental data with predictions obtained from (7.45). The material parameters α, β and γ were determined by the method outlined in Section 1C and it was found that $\alpha \equiv 0$ for the range of extension ratios considered $(1.1 \leqslant \lambda \leqslant 2.5)$. Using this result the torque M and normal force N required to maintain an extension ratio λ and angle of twist per unit length ψ are (Goldberg and Lianis, 1967)

$$M = \tfrac{1}{2}\pi\psi\lambda d^4[\beta(t) + \lambda\gamma(t)]$$

$$N = \pi d^2(\lambda - \lambda^{-2})[\beta(t) + \lambda\gamma(t)] - \tfrac{1}{4}\pi\psi^2\lambda d^4[2\beta(t) + \lambda\gamma(t)] \tag{7.46}$$

where d is the deformed radius of the cylinder. Experimental results were plotted as a series of isochronous curves of M versus ψ and N versus ψ^2 for various values of λ between 1 and 2 [values of ψ were investigated from 0 to 1.05 rad/in] and compared with the straight lines predicted by (7.46). In general the agreement between experimental data and theoretical prediction was excellent; although some data differed from the theoretical prediction by as much as 8%, this can largely be attributed to experimental difficulties and experimental scatter.

It is worth recalling that the material parameters α, β and γ were determined from data for $1.1 \leqslant \lambda \leqslant 2.5$ and are therefore not necessarily valid for $\lambda = 1$. Thus in pure torsion tests, in which $\lambda = 1$ and (7.46) requires application of a normal force

$$N_t = -\tfrac{1}{4}\pi\psi^2 d^4[2\beta(t) + \gamma(t)] \tag{7.47}$$

it is questionable whether the values of β and γ for larger extension ratios can be used. Goldberg and Lianis (1967) showed that this substitution is not valid, but that (7.47) agreed with experiment for values of β and γ determined in small strain tests. This observation is equivalent to a flattening of the curves σ_R versus λ^{-1} in simple extension in the neighbourhood of $\lambda = 1$. Consequences of this fact are detectable *a posteriori* in Goldberg and Lianis's (1967) isochrones of M versus ψ and N versus ψ^2 for values of λ near unity.

It has been shown that the Lianis model (7.36) is indistinguishable from the BKZ fluid model (7.17) under conditions of single-step stress-relaxation. Since the latter contains only three material functions, it follows that it is not possible to determine completely the four stress-relaxation functions of the Lianis theory under these conditions alone. It is necessary to consider a different form of time-dependent history in order to obtain a fourth relation connecting these functions. Each of the special cases considered in the following subsections can be used for this purpose, or can be used as a check on the validity of the Lianis model if the fourth relation has already been obtained from one of the other tests.

C. Small Uniaxial Strain Superposed on Large Equilibrium Deformation

Consider simple extension of a rod of material such that

$$x_1(\tau) = \lambda(\tau)\, X_1, \qquad x_2(\tau) = \lambda^{-1/2}(\tau)\, X_2, \qquad x_3(\tau) = \lambda^{-1/2}(\tau)\, X_3 \tag{7.48}$$

where the extension ratio $\lambda(\tau)$ is given by

$$\lambda(\tau) = \lambda_0 + \lambda_0\, \eta(\tau), \qquad -\infty \leqslant \tau \leqslant \infty \tag{7.49}$$

and λ_0 is a constant. When stresses are evaluated for times $t > 0$, the first term in (7.49) represents a large deformation for which equilibrium conditions have been attained. The second term represents a small superposed deformation; the function $\eta(\tau)$ will be assumed sufficiently small so that a theory linear in $\eta(\tau)$ can be considered.

Using (7.48) and (7.49), and linearizing with respect to η, it is readily shown that

$$\mathbf{B}(\tau) = \mathrm{diag}\{\lambda_0^2[1 + 2\eta(\tau)],\ \lambda_0^{-1}[1 - \eta(\tau)],\ \lambda_0^{-1}[1 - \eta(\tau)]\}$$

$$\mathbf{C}_t(\tau) = \mathrm{diag}\{1 + 2\eta(\tau) - 2\eta(t),\ 1 - \eta(\tau) + \eta(t),\ 1 - \eta(\tau) + \eta(t)\}. \tag{7.50}$$

When these expressions are substituted into (7.36) there results a pair of equations for $\sigma_{11} \equiv \sigma(t)$ and $\sigma_{22} \equiv 0$ from which p can be eliminated to give an expression for $\sigma(t)$. This expression is made up of an equilibrium component $\sigma_e(t)$ and a component $\sigma'(t)$ which is linear in η; this latter component is given by

$$\begin{aligned}\sigma'(t) = {} & [(a - 3b)(2\lambda_0^2 + \lambda_0^{-1}) + c(\lambda_0 + 2\lambda_0^{-2}) + b(4\lambda_0^4 + \lambda_0 + 4\lambda_0^{-2})]\, \eta(t) \\ & + 2 \int_{-\infty}^{t} [3\phi_0(t - \tau) + (2\lambda_0^2 + \lambda_0^{-1})\, \phi_1(t - \tau) + (2\lambda_0^4 + \lambda_0^{-2})\, \phi_2(t - \tau) \\ & + (\lambda_0^2 - \lambda_0^{-1})^2\, \phi_3(t - \tau)]\, \dot{\eta}(\tau)\, d\tau \end{aligned} \tag{7.51}$$

which represents the extra stress due to η.

When the extra deformation has the form of a stress-relaxation test, so that

$$\eta(t) = \eta_0\, H(t) \tag{7.52}$$

equation (7.51) yields

$$\begin{aligned}\sigma'(t)/\eta_0 = {} & [(a - 3b)(2\lambda_0^2 + \lambda_0^{-1}) + c(\lambda_0 + 2\lambda_0^{-2}) + b(4\lambda_0^4 + \lambda_0 + 4\lambda_0^{-2})] \\ & + 2[3\phi_0(t) + (2\lambda_0^2 + \lambda_0^{-1})\, \phi_1(t) + (2\lambda_0^4 + \lambda_0^{-2})\, \phi_2(t) \\ & + (\lambda_0^2 - \lambda_0^{-1})\, \phi_3(t)] \end{aligned} \tag{7.53}$$

for $t > 0$. If a test of the form represented by (7.49) and (7.52) is performed in addition to the simple relaxation tests considered in Section 1C, then (7.53) provides a fourth relation for determining the Lianis stress-relaxation functions ϕ_α.

DeHoff and Lianis (1965) performed the tests outlined above on an Estane polyurethane and an Ethylene propylene rubber. For the first material the experimental data could be modelled adequately by equations (7.22) and (7.53) and the material parameters a, b, c and ϕ_α were determined. In particular

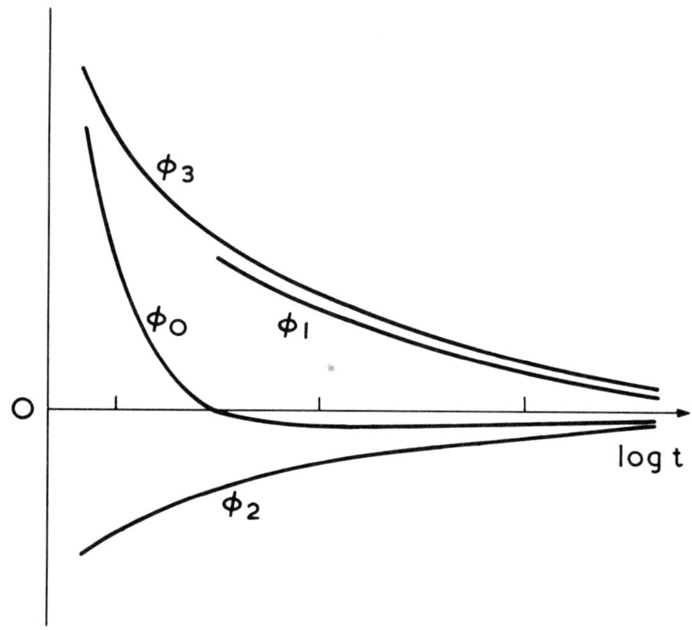

Fig. 3. Form of stress-relaxation functions for an Estane polyurethane (after DeHoff and Lianis, 1965).

$b = 0$ and the functions ϕ_α have the forms shown in Fig. 3. DeHoff and Lianis (1965) claim that the negative values of ϕ_0 are probably attributable to the fact that equilibrium conditions were only approximately reached in the time-scale of the data; the fact that ϕ_2 is negative has no special significance since it always appears in combination with positive terms.

In the case of the Ethylene propylene rubber DeHoff and Lianis (1965) found that their data could not be modelled satisfactorily by (7.22). They proposed an alternative constitutive relation involving a larger number of relaxation functions, but these could not all be determined from the available data (see also, DeHoff *et al.*, 1966).

D. Small Uniaxial Sinusoidal Oscillations Superposed on Equilibrium Deformation

When the extension ratio λ_0 is regarded as a deformed reference configuration with $\eta(\tau)$ denoting a small additional strain, a situation arises which is similar to that studied in Section 1D of Chapter 2. Applying the complex Fourier transforms (2.18) to (7.51) yields

$$\begin{aligned}\bar{\sigma}'(\omega) = {} & [(a-3b)(2\lambda_0^2+\lambda_0^{-1})+c(\lambda_0+2\lambda_0^{-2})+b(4\lambda_0^4+\lambda_0+4\lambda_0^{-2})]\,\bar{\eta}(\omega)\\ & +2[3\Phi_0(i\omega)+(2\lambda_0^2+\lambda_0^{-1})\,\Phi_1(i\omega)+(2\lambda_0^4+\lambda_0^{-2})\,\Phi_2(i\omega)\\ & +(\lambda_0^2-\lambda_0^{-1})^2\,\Phi_3(i\omega)]\,\bar{\eta}(\omega)\end{aligned} \tag{7.54}$$

where the functions

$$\Phi_\alpha(i\omega) = i\omega\,\bar{\phi}_\alpha(i\omega) \tag{7.55}$$

are complex moduli corresponding to the stress-relaxation functions ϕ_α. In terms of the complex Young's modulus $E(i\omega)$ defined by

$$\bar{\sigma}'(\omega) = E(i\omega)\,\bar{\eta}(\omega) \tag{7.56}$$

(7.54) gives

$$\begin{aligned}E_1(\omega) = {} & [(a-3b)(2\lambda_0^2+\lambda_0^{-1})+c(\lambda_0+2\lambda_0^{-2})+b(4\lambda_0^4+\lambda_0+4\lambda_0^{-2})]\\ & +2[3\Phi_{01}(\omega)+(2\lambda_0^2+\lambda_0^{-1})\,\Phi_{11}(\omega)+(2\lambda_0^4+\lambda_0^{-2})\,\Phi_{21}(\omega)\\ & +(\lambda_0^2-\lambda_0^{-1})^2\,\Phi_{31}(\omega)]\\ E_2(\omega) = {} & 2[3\Phi_{02}(\omega)+(2\lambda_0^2+\lambda_0^{-1})\,\Phi_{12}(\omega)\\ & +(2\lambda_0^4+\lambda_0^{-2})\,\Phi_{22}(\omega)+(\lambda_0^2-\lambda_0^{-1})^2\,\Phi_{32}(\omega)]\end{aligned} \tag{7.57}$$

where

$$E(i\omega) = E_1(\omega)+iE_2(\omega), \qquad \Phi_\alpha(i\omega) = \Phi_{\alpha 1}(\omega)+i\Phi_{\alpha 2}(\omega) \tag{7.58}$$

and E_1, E_2, $\Phi_{\alpha 1}$ and $\Phi_{\alpha 2}$ are real functions of ω.

Equation (7.56) is similar in form to equation (2.20) for linear one-dimensional behaviour; it differs only in a suppressed dependence in (7.56) of $E(i\omega)$ on λ_0. Expressions (7.57) show how this complex modulus depends on λ_0 and Lianis's relaxation functions.

It is interesting to compare the predictions of the Lianis theory with those of the BKZ fluid theory. When (7.48) and (7.49) are substituted into (7.12) it is readily shown that

$$\mathbf{B}_\tau = \operatorname{diag}\{1+2\eta(t)-2\eta(\tau),\ 1+\eta(\tau)-\eta(t),\ 1+\eta(\tau)-\eta(t)\} \tag{7.59}$$

where powers of η have been neglected. Equation (7.17) now yields

$$\sigma(t) = 3 \int_{-\infty}^{t} [m(t-\tau) + b(t-\tau)]\,[\eta(t) - \eta(\tau)]\,d\tau \tag{7.60}$$

where $\sigma_{11} = \sigma(t)$ and $\sigma_{22} = 0$. Using (7.19) and integration by parts this can be written as

$$\sigma(t) = 3 \int_{-\infty}^{t} [M(t-\tau) + B(t-\tau)]\,\dot{\eta}(\tau)\,d\tau \tag{7.61}$$

which is equivalent to

$$\sigma(t) = 3 \int_{-\infty}^{t} [\beta(t-\tau) + \gamma(t-\tau)]\,\dot{\eta}(\tau)\,\mathrm{d}\tau. \tag{7.62}$$

Equation (7.62) shows that the BKZ theory predicts no equilibrium component $\sigma_e(t)$ so that $\sigma(t) = \sigma'(t)$, and this is independent of λ_0. In the notation of (7.56)–(7.58) the BKZ fluid theory predicts that

$$E_1(\omega) = -3\omega[\bar{\beta}_2(\omega) + \bar{\gamma}_2(\omega)], \qquad E_2(\omega) = 3\omega[\bar{\beta}_1(\omega) + \bar{\gamma}_1(\omega)] \tag{7.63}$$

where

$$\bar{\beta}(\omega) = \bar{\beta}_1(\omega) + i\bar{\beta}_2(\omega), \qquad \bar{\gamma}(\omega) = \bar{\gamma}_1(\omega) + i\bar{\gamma}_2(\omega) \tag{7.64}$$

and $\bar{\beta}_1$, $\bar{\beta}_2$, $\bar{\gamma}_1$ and $\bar{\gamma}_2$ are real functions of ω.

It is evident from the above discussion that the prediction of the BKZ fluid model is inadequate for describing the behaviour of real materials in the present situation. In order to remedy this Goldberg and Lianis (1968) proposed a modified BKZ theory in which the finite strain λ_0 is considered to be applied at time $-T_0$ where T_0 is finite (it is infinite in (7.49) *et seq*); this results in an equilibrium stress for $t > 0$ if it is further assumed that $\alpha(t) = \alpha(T_0)$, $\beta(t) = \beta(T_0)$, $\gamma(t) = \gamma(T_0)$ for $t \geqslant T_0$. The effect of these assumptions on (7.63) is to introduce an elastic component into $E_1(\omega)$ of the same form as that contained in (7.57) where

$$a - 2b = \gamma(T_0), \qquad 2b + c = \beta(T_0), \qquad b = \alpha(T_0) \tag{7.65}$$

and, in addition, to replace the functions β and γ in (7.63) by $\Delta\beta$ and $\Delta\gamma$ defined by

$$\Delta\beta(t) = \beta(t) - \beta(T_0), \qquad \Delta\gamma(t) = \gamma(t) - \gamma(T_0). \tag{7.66}$$

The function $E_2(\omega)$ and the non-elastic part of $E_1(\omega)$ remain independent of λ_0.

Goldberg and Lianis (1968) made an experimental investigation of the validity of their prediction (7.57) and that of their modified BKZ theory. If tests are performed for a particular value λ_0, then (2.20) and (7.56) are identical in form, and oscillatory tests in which η has the form

$$\eta = \eta_0 \exp(i\Omega t) \tag{7.67}$$

can be used to determine $E(i\Omega)$ for a range of values of Ω from measurements of force amplitude and phase angle. If F_e denotes the equilibrium force in a rod due to the constant extension ratio λ_0, and F' denotes the extra force due to the superposed vibration, then

$$F(t) = A_0\,\sigma(t)\,\lambda^{-1}(t) = A_0\,\lambda_0^{-1}\{\sigma_e + \eta_0[E_1(\Omega) + iE_2(\Omega)]\,e^{i\Omega t}\}\,(1 - \eta_0\,e^{i\Omega t}) \tag{7.68}$$

so that

$$F_e = A_0\,\lambda_0^{-1}\,\sigma_e$$

$$F' = A_0\,\lambda_0^{-1}\,\eta_0\,e^{i\Omega t}\{[E_1(\Omega) - \sigma_e] + iE_2(\Omega)\} \tag{7.69}$$

where A_0 is the initial cross-sectional area of the rod, and σ_e is given by

$$\sigma_e = (\lambda_0^2 - \lambda_0^{-1})\,[a + b(\lambda_0^2 + 2\lambda_0^{-1} - 3) + c\lambda_0^{-1}]. \tag{7.70}$$

From (7.69) the amplitude F_0 and phase θ of F' can be expressed in the forms

$$F_0 = A_0\,\lambda_0^{-1}\,\eta_0\{[E_1(\Omega) - \sigma_e]^2 + [E_2(\Omega)]^2\}^{1/2}$$

$$\tan\theta = E_2(\Omega)\{E_1(\Omega) - \sigma_e\}^{-1} \tag{7.71}$$

and measurement of these two quantities is sufficient to determine $E_1(\Omega)$ and $E_2(\Omega)$.

Goldberg and Lianis (1968) compared the values of E_1 and E_2 calculated from (7.71) and their own experimental data with the values calculated from (7.57) and the modified form of BKZ theory using functions ϕ_α, β and γ determined from other tests. They experimented with specimens of an SBR rubber, Thiokol crosslinked gum stock, and Estane $\times$100 un-crosslinked polyurethane, in the frequency range 0.01 to 0.5 Hz and extension ratios λ_0 in the range $1.5 \leqslant \lambda_0 \leqslant 2.0$. The average strain amplitude during the tests was 0.05. A low frequency range was used because of lack of accuracy of stress-relaxation data for times less than 1 sec, and because inertia effects were considered to be significant at higher frequencies. In general Goldberg and Lianis (1968) found that the Lianis theory gave significantly better agreement with experimental data than did the modified BKZ theory. Although this agreement was adequate in some cases, notably for the real part of the modulus for the Thiokol and SBR materials, some predictions were in error by as much as 50%. The authors attribute this to heat generation in the specimen; whatever the source of error it is evident that there is room for improvement of the proposed models for these dynamic situations.

E. Uniaxial Constant Stretch Rate

McGuirt and Lianis (1967a) compared the theoretical predictions of the Lianis theory with experimental observations for constant stretch rate loading and for constant stretch rate unloading. In the former case

$$\lambda(\tau) = 1 + R_1 \tau, \qquad 0 \leqslant \tau \leqslant t \tag{7.72}$$

where R_1 is a positive constant, and in the latter case

$$\begin{aligned} \lambda(\tau) &= 1 + (\lambda_1 - 1) H(\tau), \qquad -\infty < \tau \leqslant t_0 \\ \lambda(\tau) &= \lambda_1 + R_2(\tau - t_0), \qquad t_0 \leqslant \tau \leqslant t \end{aligned} \tag{7.73}$$

where R_2 is a negative constant. McGuirt and Lianis (1967a) studied a Thiokol sheetstock material and two SBR materials. In the loading tests R_1 ranged between approximately 4×10^{-3} sec^{-1} and 120×10^{-3} sec^{-1}, and data were taken at times for which λ did not exceed 2.2. In the unloading tests λ_1 was approximately 2 and R_2 ranged between -5×10^{-3} sec^{-1} and -30×10^{-3} sec^{-1}; data were recorded until λ had decreased to approximately 1.45. The agreement between experimental data and theoretical prediction was particularly good for the loading tests and quite good for the unloading tests. In the latter case the theory tends to underestimate the required stress, but was never in error by more than 8%.

F. Uniaxial Two-Step Relaxation

A particularly simple test which Lianis and his co-workers have used in conjunction with single-step relaxation tests to determine the relaxation functions is the two-step relaxation test. In this case

$$\lambda(\tau) = 1 + (\lambda_1 - 1) H(\tau) + (\lambda_2 - \lambda_1) H(\tau - t_1) \qquad -\infty < \tau \leqslant t. \tag{7.74}$$

If t_1 is sufficiently large for the stress to have reached equilibrium with respect to the extension ratio λ_1, then (7.36) gives the stress required to maintain the deformation (7.74) in the form

$$\begin{aligned} \sigma(t') = \sigma_e(\lambda_2) &+ 2[\lambda_2 \lambda_1^{-1} - \lambda_1^2 \lambda_2^{-2}] \phi_0(t') + 2[\lambda_2^2 - \lambda_1^2 - \lambda_2^{-1} + \lambda_1^{-1}] \phi_1(t') \\ &+ 2[\lambda_2^4 - \lambda_1^2 \lambda_2^2 - \lambda_2^{-2} + \lambda_1^{-1} \lambda_2^{-1}] \phi_2(t') \\ &+ [\lambda_2^4 + \lambda_2 - \lambda_2^2 \lambda_1^2 - 2\lambda_2^2 \lambda_1^{-1} - 2\lambda_2^{-2} + \lambda_1^2 \lambda_2^{-1} + 2\lambda_1^{-1} \lambda_2^{-1}] \phi_3(t') \end{aligned} \tag{7.75}$$

where $t' = t - t_1 > 0$ and $\sigma_e(\lambda_2)$ represents the equilibrium stress corresponding to the extension ratio λ_2. Equation (7.75) provides a further relation connecting the ϕ_α which can be used in the determination of these functions.

G. Exponential Extension Ratio History

An exponential extension ratio history is defined by

$$\lambda(\tau) = 1, \qquad \tau \leqslant 0$$
$$\lambda(\tau) = e^{K\tau}, \qquad \tau \geqslant 0 \tag{7.76}$$

and it is a simple matter to calculate the stresses corresponding to this history as predicted by the BKZ theory (7.17) and the Lianis theory (7.36). When the material functions are already known from other tests, these predictions can be compared with the experimental data from experiments in which (7.76) is applied. Such a comparison has been made by Goldberg *et al.* (1966) for styrene-butadiene rubber, polyisobutylene and polyurethane rubbers. It was found that both theories gave accurate predictions for the SBR and PIB materials; exponential parameters K were considered in the range 0.0051 to 0.0425 and the maximum extension ratio attained was approximately 3.6. For the polyurethane rubbers, and similar values of K, the theories could not adequately describe the experimental data and, at the largest times considered, the theories gave predictions with errors approaching 50%.

3. MODIFIED FORM OF LIANIS'S THEORY

A. Summary of Experimental Tests of Lianis's Theory

With the exception of the combined tension-torsion tests of Goldberg and Lianis (1967) all of the experimental data discussed so far in this chapter refer to uniaxial extension tests. This single non-uniaxial investigation was conducted under single-step stress-relaxation conditions and cannot, therefore, give more than very incomplete information about non-uniaxial behaviour.

As far as uniaxial data are concerned, the BKZ and Lianis theories have had a good degree of success, but there have been exceptions either for particular materials or for particular forms of time-dependent strain history. For almost all of the materials investigated both theories were able to model the experimental data for uniaxial single or multiple-step stress-relaxation. However, in the case of DeHoff and Lianis's (1965) data for an Ethylene propylene rubber the theories were found to be too restrictive and these authors proposed an alternative model for this material involving a larger number of relaxation functions. For other forms of time-dependent history the Lianis theory, and in some cases the BKZ theory also, was adequate for at least some materials. The only noticeable trend in these results appears to be that the Lianis theory seems to be more successful in making predictions for materials for which either

$\alpha(t) \equiv 0$ all t, or $\alpha(\infty) = 0$. From (7.44) it is seen that this condition implies that $b \equiv 0$, and it follows that successful correlations between theory and experiment are not a good test of the assumptions (7.34). Since derivation of the Lianis theory was based upon the assumption of a particular form of long-term behaviour, it follows that uniaxial experiments are not a strong test of the theory; more general non-uniaxial experiments are required for this purpose.

B. The McGuirt-Lianis Theory

As a further investigation of the Lianis theory, McGuirt and Lianis (1967b) considered the theoretical predictions for the case of equal biaxial single-step stress-relaxation. For this situation

$$x_1(\tau) = \lambda(\tau)\, X_1, \qquad x_2(\tau) = \lambda(\tau)\, X_2, \qquad x_3(\tau) = \lambda^{-2}(\tau)\, X_3 \tag{7.77}$$

where

$$\lambda(\tau) = 1 + (\lambda_0 - 1)\, H(\tau). \tag{7.78}$$

The Lianis theory (7.36) now predicts that the equilibrium stresses are

$$\begin{aligned} \sigma^{(e)} = \sigma_{11}^{(e)} = \sigma_{22}^{(e)} &= -p + [a - 3b + (b + c)(2\lambda_0^2 + \lambda_0^{-4})]\,\lambda_0^2 - c\lambda_0^4 \\ 0 = \sigma_{33}^{(e)} &= -p + [a - 3b + (b + c)(2\lambda_0^2 + \lambda_0^{-4})]\,\lambda_0^{-4} - c\lambda_0^{-8} \end{aligned} \tag{7.79}$$

and upon elimination of p

$$\sigma_B^{(e)} \equiv \frac{\sigma^{(e)}}{\lambda_0^2 - \lambda_0^{-4}} = a + b[\lambda_0^{-4} + 2\lambda_0^2 - 3] + c\lambda_0^2 \tag{7.80}$$

where σ_B denotes a reduced stress for the biaxial deformation.

In a uniaxial single-step stress-relaxation test, (7.22) and (7.44) give the equilibrium stress in the form

$$\sigma_R^{(e)} \equiv \frac{\sigma_e}{\lambda_0^2 - \lambda_0^{-1}} = a + b[\lambda_0^2 + 2\lambda_0^{-1} - 3] + c\lambda_0^{-1}. \tag{7.81}$$

For the materials successfully modelled in uniaxial tests by the Lianis theory, it has been noted that $b \equiv 0$. Thus (7.80) and (7.81) predict that

$$\begin{aligned} \sigma_B^{(e)} &= a + c\lambda_0^2 \\ \sigma_R^{(e)} &= a + c\lambda_0^{-1} \end{aligned} \tag{7.82}$$

and it follows that the long-time isochrones of $\sigma_B^{(e)}$ against λ_0^2 and $\sigma_R^{(e)}$ against λ_0^{-1} should be straight lines with identical slopes and intercepts. Experiments performed by McGuirt and Lianis (1967b) on their SBR material showed that both isochrones were approximately straight lines in the experimental range of λ_0, but they did not have equal slopes and intercepts. They concluded that the Lianis theory is not adequate to model *both* uniaxial and biaxial tests, even for the SBR material which had been modelled adequately in uniaxial tests; in order to model both types of test a modified theory is required. By inspection of their equilibrium data they were led to propose the special forms

$$\phi = a + b(I_2 - 2)^{-2}$$
$$\psi = c + d(I_2 - 3). \tag{7.83}$$

Further study of the uniaxial and biaxial isochrones for earlier times then suggested a modified constitutive relation of the form

$$\begin{aligned}\boldsymbol{\sigma}(t) = -p\mathbf{I} &+ \{a + b(I_2 - 2)^{-2} + cI_1 + dI_1(I_2 - 3)\}\,\mathbf{B} \\ &- \{c + d(I_2 - 3)\}\,\mathbf{B}^2 + 2\int_{-\infty}^{t} \phi_0(t-\tau)\,\dot{\mathbf{C}}_t(\tau)\,d\tau \\ &+ 2(I_2 - 3)\int_{-\infty}^{t} \Phi_0(t-\tau)\,\dot{\mathbf{C}}_t(\tau)\,d\tau \\ &+ \int_{-\infty}^{t} \phi_1(t-\tau)\,[\mathbf{B}\dot{\mathbf{C}}_t(\tau) + \dot{\mathbf{C}}_t(\tau)\,\mathbf{B}]\,d\tau.\end{aligned} \tag{7.84}$$

In the McGuirt-Lianis theory the material is characterized by four constants a, b, c and d, and *three* stress-relaxation functions ϕ_0, Φ_0 and ϕ_1 which are functions of t alone. The particular form (7.84) arose through expansion of the relaxation functions in (7.33) as

$$\phi_\alpha(t, I_1, I_2) = \phi_\alpha(t) + (I_2 - 3)\,\Phi_\alpha(t)$$
$$\phi_{\alpha\beta}(t, I_1, I_2) = \phi_{\alpha\beta}(t) + (I_2 - 3)\,\Phi_{\alpha\beta}(t) \tag{7.85}$$

followed by neglect of many terms. Although the McGuirt and Lianis (1967b) experimental data suggested that this reduction was plausible, it is by no means rigorous or unique.

C. Experimental Determination of Material Functions

Using (7.84) it is readily shown that the stress in a single-step uniaxial stress-relaxation test is given by

$$\sigma_R(t) = \gamma(t,\ I_2) + \beta(t,\ I_2)\ \lambda_0^{-1} \tag{7.86}$$

where

$$\gamma(t, I_2) = a + b(I_2 - 2)^{-2} + 2\phi_1(t)$$
$$\beta(t, I_2) = c + 2\phi_0(t) + [d + 2\Phi_0(t)](I_2 - 3) \tag{7.87}$$

and σ_R is defined in (7.23). The invariant I_2 is given by

$$I_2 = 2\lambda_0 + \lambda_0^{-2}. \tag{7.88}$$

The expression for the stress in a biaxial test is given by

$$\sigma_B(t) = \sigma(t)[\lambda_0^2 - \lambda_0^{-4}]^{-1} = \gamma(t, I_2) + \beta(t, I_2)\lambda_0^2 \tag{7.89}$$

where γ and β are again defined by (7.87), but now I_2 is given by

$$I_2 = \lambda_0^4 + 2\lambda_0^{-2}. \tag{7.90}$$

The material functions may be determined in the following manner. For a particular time t a pair of values of σ_R and λ_0 can be read off the uniaxial isochrone (7.86). Expression (7.88) gives the value of I_2 corresponding to these values and (7.90) then gives the biaxial extension ratio corresponding to this value of I_2. The biaxial isochrone (7.89) for the same time now yields the value of σ_B corresponding to the biaxial extension ratio. This procedure gives a uniaxial and a biaxial extension ratio and their corresponding σ_R and σ_B for identical values of time and the invariant I_2; equations (7.86) and (7.89) can now be solved for $\gamma(t, I_2)$ and $\beta(t, I_2)$. If this procedure is repeated for other values of I_2 the material functions may be readily determined from isochronous plots of γ versus $(I_2 - 2)^{-2}$ and β versus $(I_2 - 3)$.

D. Predictions for Other Time-Dependent Histories

Using data from uniaxial and biaxial single-step relaxation tests in the manner described in Sub-section C, McGuirt and Lianis (1967b) were able to use (7.84) to predict *uniaxial* behaviour in constant stretch-rate loading, exponential loading, and constant stretch-rate unloading tests. The material studied was their SBR material; the experimental data of McGuirt and Lianis (1967a) and Goldberg *et al.* (1966) for this material have already been discussed in Section 2 and the form of the McGuirt-Lianis relaxation functions is shown in Fig. 4. McGuirt and Lianis (1967b) showed that the constitutive relation (7.84) can model these data as accurately as the Lianis theory, and the former has the advantage that it is partly derived from biaxial tests.

Further experimental tests of the McGuirt-Lianis theory (7.84) have been made by McGuirt and Lianis (1967c) and by Flowers and Lianis (1969), the material again being the SBR material. Uniaxial and equal biaxial tests were performed for constant and logarithmic stretch-rates: non-homogeneous biaxial tests were performed in single and double-step relaxation and in constant stretch-rate. In all of these tests the agreement between experimental data and the predictions of the theory is satisfactory. It therefore appears that the McGuirt-Lianis theory is an adequate model for SBR under the uniaxial and biaxial conditions considered. This is a notable achievement, although it should be remembered that it has been proved for only one material, and the

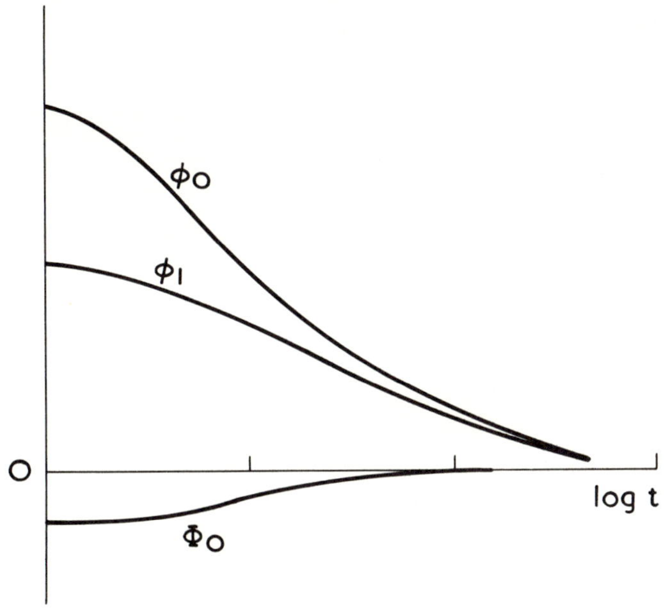

Fig. 4. Form of stress-relaxation functions for an SBR material (after McGuirt and Lianis, 1967b).

theory may well need further modification when other materials or more complex three-dimensional deformation histories are considered.

4. NON-ISOTHERMAL BEHAVIOUR

The experimental data of Morgan and Ward (1971) which were discussed in Section 3E, Chapter 5, suggest that the concept of thermo-rheologically simple behaviour can be extended to nonlinear response. In particular, their results indicate that the material functions occurring in the multiple-integral theory can be evaluated at different temperatures by using a simple shift law on logarithmic-time axes. In each of their experiments isothermal conditions prevailed, and the results show that a thermo-rheologically simple model is adequate to relate isothermal conditions at one temperature with similar conditions at other temperatures within the experimental range. It does not necessarily follow that this model can be used to predict material response in mechanical tests for which the temperature varies during the experiment.

An investigation which supports the validity of time-temperature equivalence for variable temperature conditions has been reported by McGuirt and Lianis (1968). Experimental data were obtained within the temperature range −30°C to 27°C for the SBR material investigated in their earlier experiments (see previous sections of this chapter). Theoretical modelling of material behaviour and prediction of response in non-isothermal conditions was achieved through a generalization of the isothermal theories discussed earlier in this chapter. In particular, the non-isothermal form of the Lianis theory (7.36) is

$$\begin{aligned}\boldsymbol{\sigma}(t) = -p(t)\mathbf{I} + [a + b(I_1 - 3) + cI_1]\mathbf{B} - c\mathbf{B}^2 \\ + 2\int_{-\infty}^{t} \phi_0(\xi - \xi')\dot{\mathbf{C}}_t(\tau)\,d\tau + \int_{-\infty}^{t} \phi_1(\xi - \xi')[\mathbf{B}\dot{\mathbf{C}}_t(\tau) + \dot{\mathbf{C}}_t(\tau)\mathbf{B}]\,d\tau \\ + \int_{-\infty}^{t} \phi_2(\xi - \xi')[\mathbf{B}^2\dot{\mathbf{C}}_t(\tau) + \dot{\mathbf{C}}_t(\tau)\mathbf{B}^2]\,d\tau + \mathbf{B}\int_{-\infty}^{t} \phi_3(\xi - \xi')\dot{I}_1(\tau)\,d\tau\end{aligned} \tag{7.91}$$

where $\mathbf{B} = \mathbf{B}(t)$ and ξ and ξ' are the reduced times defined in (3.37). Constitutive equations (7.36) and (7.91) differ only in so far as the material functions in the non-isothermal form (7.91) are evaluated at the reduced time. For isothermal conditions the reduced time is given by (3.38) and, at the reference temperature, reduced and real time are identical and (7.91) is identical with (7.36).

In a uniaxial single-step stress-relaxation test equation (7.91) predicts a response

$$\sigma_R(t) = (\lambda^2 - 1)\,\alpha(\xi) + \lambda^{-1}\beta(\xi) + \gamma(\xi) \tag{7.92}$$

where σ_R is defined in (7.23) and α, β and γ are defined in (7.44). McGuirt and Lianis (1968) performed a series of isothermal single-step relaxation tests at various temperatures in the range −30°C to 27°C. In each case they found that the isochrones of σ_R versus λ^{-1} were adequately represented by straight lines, so that a good approximation is obtained by setting

$$\alpha(\xi) = 0, \qquad b = 0, \qquad \phi_3(\xi) = -2\phi_2(\xi). \tag{7.93}$$

In addition it was noted that all isochrones had the same intercept, within the limits of experimental accuracy, so that

$$\phi_1(\xi) = \phi_3(\xi) \tag{7.94}$$

and (7.92) assumes the simplified form

$$\sigma_R(t) = a + \lambda^{-1}\beta(\xi) \tag{7.95}$$

where

$$\beta(\xi) = c + 2\phi_0(\xi) + \phi_3(\xi). \tag{7.96}$$

It should be noted that the assumption of time-temperature equivalence does not permit the constants a and c to be temperature-dependent.

For the isothermal conditions considered, ξ is proportional to t, the constant of proportionality being the shift factor $a(T)$. Thus, if the material behaviour satisfies time-temperature equivalence, curves of β versus $\log t$ obtained at different temperatures should be capable of shifting to form a master curve for the reference temperature, and the amount of shift determines the shift factor $a(T)$. The data of McGuirt and Lianis (1968) for SBR satisfied this requirement within the range of experimental data; the master curve spanned five decades of time. Although the measurements of the shift functions were based on finite strain data, the results agree with previous measurements obtained for infinitesimal strain. This result is similar to that obtained by Morgan and Ward (1971) for polypropylene (see Chapter 5, Section 3E).

When the master curve for β and the shift function are known from isothermal tests at different temperatures, equation (7.95) can be used to predict uniaxial single-step relaxation behaviour under variable temperature conditions. For a known temperature history σ_R is given by (7.95) in which ξ is determined from (3.37). A typical comparison between theoretical prediction and experimental data is shown in Fig. 5. These results were obtained by McGuirt and Lianis (1968) for an extension ratio of $\lambda = 1.727$ in the relaxation test and a temperature variation from $-30°C$ to $15°C$ during the course of the experiment. It is evident that there is good agreement between experimental data and theoretical prediction. Also shown in Fig. 5 are isothermal single-step relaxation curves for $-30°C$ and $15°C$; it can be seen that the variable-temperature curve makes a transition between these curves as might be expected.

McGuirt and Lianis (1968) also considered the combined effect of constant stretch-rate uniaxial loading and variable temperature conditions. For this loading history the material cannot be characterized entirely through the combination of relaxation functions (7.96) and an additional test, such as double-step relaxation, is required in order to determine separate values for the ϕ_α. However, since the shift function was already known, it was necessary to obtain the double-step data only at $-30°C$ and $0°C$; this was sufficient to form the required master curves. For the material considered (SBR) it was found that the approximation $\phi_0 = \phi_1$ could be used with sufficient accuracy, and the predicted stresses in constant stretch-rate tests then showed good agreement with experimental data when the temperature was changed from $-30°C$ to $15°C$ during the course of the experiment.

As noted in Section 3B the Lianis theory (7.36) is not adequate to model biaxial conditions, and a modified equation must be considered. McGuirt and

Lianis (1968) employed the non-isothermal form of a modified constitutive relation proposed by McGuirt and Lianis (1967d). For the new equation, which differs in some respects from (7.84), it was demonstrated that thermo-rheologically simple behaviour was applicable to equal biaxial relaxation data in addition to the previous uniaxial data for the variable temperature histories considered.

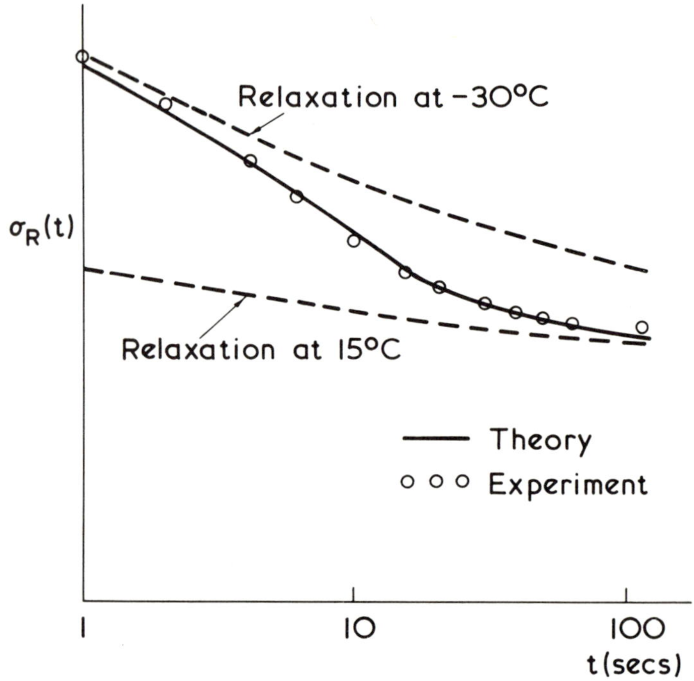

Fig. 5. Single-step stress-relaxation of SBR with temperature varying from −30°C to 15°C (after McGuirt and Lianis, 1968).

From the results of McGuirt and Lianis (1968) and Morgan and Ward (1971) there is a strong indication that the concept of thermo-rheologically simple behaviour is a useful model for nonlinear behaviour, at least for the materials and range of conditions considered. It is noted that most of these results refer to uniaxial behaviour, and the modifications which have been made to the Lianis theory have been occasioned by the introduction of multi-axial behaviour and not by the temperature variation. Present evidence therefore suggests that it may not be difficult to model non-isothermal behaviour of a wide class of materials once a satisfactory nonlinear model for isothermal behaviour has been obtained.

Chapter 8 Material Characterization in Special Situations

It is evident from the discussion of previous chapters that characterization of nonlinear viscoelastic behaviour is a difficult and complex task if one wishes to model a wide class of materials under relatively general conditions. For some purposes this may be desirable, or even essential, but there are also a considerable number of applications for which a restricted form of characterization is adequate. This restriction may arise through interest in a class of materials whose behaviour is in some sense degenerate, or it may occur because only a particular class of deformations or loadings is to be considered. Some examples of this have been described in previous chapters where, apart from the assumed restriction, the theories are quite general. In particular, a theory for one-dimensional behaviour was considered in Chapter 5, and theories for plane stress and plane strain were discussed in Chapter 6. These theories refer to special classes of deformation. Apart from the assumptions of isotropic and isothermal behaviour, which are made in the bulk of the preceding discussion, the only restricted form of material behaviour which has been considered is that for incompressible materials.

In this chapter some further restricted theories are added to those mentioned above. In some cases the particular restriction has already been discussed, but in combination with other assumptions. Here the effects of each restriction are treated in isolation from the other assumptions.

A situation which arises frequently in practice, at least to a good approximation, is that where the material is subjected to single-step stress-relaxation. This occurs when a plastics component is deformed into a new configuration during assembly of a structure containing more rigid elements. Examples are plastics components which are bolted into metal structures, and various forms of seals which are deformed during assembly and are then left for considerable periods without extra tightening. Single-step stress-relaxation has been considered previously in the context of the BKZ and Lianis theories, but these discussions depend also on the special constitutive relations. In Section 1 single-step stress-relaxation is considered in a more general context, and the dual case of single-step creep is treated in Section 2. The latter situation is relevant to a wide class of practical applications where structures are subjected to loading situations which are essentially time-independent once they have been applied.

In such practical applications the applied load or deformation is not exactly of the form assumed in single-step tests since some finite time is required for the load or deformation to be applied. However, if the time scale of interest is large compared with this time, the step-loading is an adequate approximation. Thus, the performance of a seal after a period of many days is unlikely to depend on the precise nature of the time-history of deformation during the few seconds required to tighten the seal initially. In this sense the applications envisaged for single-step modelling are long-term predictions. When short-term response is of practical interest, where the characteristic time of the loading history is not small compared with the time of interest, a theory for short-term behaviour may be appropriate; theories of this type are considered in Section 3.

As noted above, many practical applications involve loading or deformation histories which are essentially time-independent after the initial application. If deviations from time-independence can be neglected, then the special theories for single-step histories are appropriate. However, if the deviations are small but significant, it is necessary to consider a theory for small deviations superposed on a large constant history: theories of this type are discussed in Section 4.

All of the special situations discussed above refer to some assumption regarding the form of the deformation or loading history although, to be more precise, the material properties govern the range of conditions under which the theories are valid. In Section 5 a theory is proposed which relates to materials whose reponse is almost elastic; this may be of practical value where an elastic model is adequate as a first approximation, but more precise modelling is required.

1. SINGLE-STEP STRESS-RELAXATION IN GENERAL DEFORMATION

A. Constitutive Relation

The origin of the time scale is chosen as the instant of time when the deformation is applied: the coordinates of a typical material element are denoted by a_i for times $\tau < 0$, and by b_i for times $\tau > 0$ with a_i and b_i time-independent. If the material has a preferred reference state, then it is assumed to be in that configuration for $\tau < 0$, so that $a_i \equiv X_i$. When $\tau < 0$ the deformation gradient matrix $\mathbf{F}$ is the unit matrix $\mathbf{I}$; when $\tau > 0$ it is denoted by $\mathbf{F}_0$, where

$$\mathbf{F}_0 = \|\partial b_i/\partial a_j\|. \tag{8.1}$$

It follows that

$$\mathbf{F}(\tau) = \mathbf{I} + (\mathbf{F}_0 - \mathbf{I})\, H(\tau) \tag{8.2}$$

and

$$\mathbf{C}(\tau) = \mathbf{I} + (\mathbf{C}_0 - \mathbf{I})\, H(\tau), \qquad \mathbf{C}_0 = \mathbf{F}_0^T \mathbf{F}_0 \tag{8.3}$$

where $\mathbf{F}_0$ and $\mathbf{C}_0$ are independent of time. The functional in (4.34) can now be written in the form

$$\underset{\tau=-\infty}{\overset{t}{\mathscr{F}^*}} \{\mathbf{C}(\tau)\} = \mathbf{f}(\mathbf{C}_0, t) \tag{8.4}$$

where $\mathbf{f}$ is a matrix function which, for an isotropic material, becomes

$$\mathbf{f}(\mathbf{C}_0, t) = \beta_0 \mathbf{I} + \beta_1 \mathbf{C}_0 + \beta_2 \mathbf{C}_0^2 \tag{8.5}$$

where the β_i are functions of time and of the invariants (4.46) evaluated with $\mathbf{C}_0$ replacing $\mathbf{B}$. For times $t > 0$ the general constitutive relation (4.34) for an isotropic material subjected to single-step stress-relaxation now takes the form

$$\begin{aligned} \boldsymbol{\sigma}(t) &= \mathbf{F}_0(\beta_0 \mathbf{I} + \beta_1 \mathbf{C}_0 + \beta_2 \mathbf{C}_0^2)\mathbf{F}_0^T \\ &= \beta_0 \mathbf{B}_0 + \beta_1 \mathbf{B}_0^2 + \beta_2 \mathbf{B}_0^3 \end{aligned} \tag{8.6}$$

where $\mathbf{B}_0$ is the left Cauchy-Green strain matrix evaluated using $\mathbf{F}_0$. The Hamilton-Cayley Theorem (7.41) can now be used to reduce (8.6) to

$$\boldsymbol{\sigma}(t) = \alpha_0 \mathbf{I} + \alpha_1 \mathbf{B}_0 + \alpha_2 \mathbf{B}_0^2 \tag{8.7}$$

where the α_i are functions of time and of the invariants of $\mathbf{B}_0$. Equation (8.7) differs from the elastic constitutive equation (4.45) only in so far as the functions α_i contain the extra dependence on time.

If the material has no preferred configuration a constitutive equation of the form (4.20) is a more appropriate starting point. In this case it follows that, for times $t > 0$, $\mathbf{F}_t(\tau) = \mathbf{I}$ when $\tau > 0$ and $\mathbf{F}_t(\tau) = \|\partial a_i/\partial b_j\|$ when $\tau < 0$. Thus $\mathbf{C}_t(\tau)$ can be written

$$\mathbf{C}_t(\tau) = \mathbf{B}_0^{-1} + (\mathbf{I} - \mathbf{B}_0^{-1})\, H(\tau) \tag{8.8}$$

and (4.20) reduces to

$$\boldsymbol{\sigma}(t) = \mathbf{g}(\mathbf{B}_0, t) \tag{8.9}$$

where $\mathbf{g}$ is an isotropic matrix function. The explicit form of (8.9) is (8.7), so the two approaches yield the same constitutive relation, provided that $\mathbf{B}_0$ is evaluated using (8.1) and a_i is identified with the natural reference state if one exists.

When the material is represented adequately by the assumption of incompressibility (8.7) is replaced by (4.49) in which α_1 and α_2 depend on time in addition to the invariants I_1 and I_2 which are evaluated for $\mathbf{B} \equiv \mathbf{B}_0$. For conditions of plane strain $\mathbf{B}_0$ is essentially two-dimensional and (8.7) is replaced by

$$\bar{\boldsymbol{\sigma}}(t) = \alpha_0 \bar{\mathbf{I}} + \alpha_1 \bar{\mathbf{B}}_0 \tag{8.10}$$

where the bar denotes a 2×2 matrix; α_0 and α_1 are functions of t and of the invariants $\operatorname{tr} \bar{\mathbf{B}}_0$ and $\operatorname{tr} \bar{\mathbf{B}}_0^2$ (see Rivlin, 1955).

B. Small Deformations

An equation of the form (8.7) was used by Bergen *et al.* (1960) to describe small deformations of an incompressible isotropic material. They assumed that

$$\mathbf{B} \simeq \mathbf{I} + 2\boldsymbol{\epsilon} \tag{8.11}$$

for small deformations, where $\boldsymbol{\epsilon}$ is the classical strain measure (1.8) and the suffix zero is omitted from this and subsequent expressions. When (8.11) is substituted into the incompressible form of (8.7) there results

$$\boldsymbol{\sigma}(t) = -p\mathbf{I} + \Theta\boldsymbol{\epsilon} + \Phi\boldsymbol{\epsilon}^2 \tag{8.12}$$

where Θ and Φ are functions of t and of the invariants which can be expressed in the convenient forms

$$J_1 = 2[\operatorname{tr} \boldsymbol{\epsilon}^2 - (\operatorname{tr} \boldsymbol{\epsilon})^2], \qquad J_2 = 8 \det \boldsymbol{\epsilon}. \tag{8.13}$$

Since Θ and Φ depend on the invariants, the magnitudes of these functions depend on the magnitude of $\boldsymbol{\epsilon}$. Bergen *et al.* (1960) assumed that the condition $\Phi \gg \Theta$ is not valid, so that (8.12) can be approximated by

$$\boldsymbol{\sigma}(t) = -p\mathbf{I} + \Theta\boldsymbol{\epsilon} \tag{8.14}$$

and the dependence of Θ on its arguments is to be determined from experimental data. For this purpose Bergen *et al.* (1960) considered simultaneous extension and torsion of a circular tube, and showed that the moment M and normal force N can be expressed as

$$M = \pi\psi \int_a^b \Theta r^3 \, dr, \qquad N = 3\pi e \int_a^b \Theta r \, dr \tag{8.15}$$

where e is the extensional strain, ψ is the angle of twist per unit length of extended tube, and a and b are the internal and external radii of the tube, respectively. The invariants are given by

$$J_1 = 3e^2 + \psi^2 r^2, \qquad J_2 = 2e^3 + e\psi^2 r^2 \tag{8.16}$$

where r denotes the radial coordinate; thus Θ is position dependent.

If the tube is sufficiently thin-walled and Θ does not vary rapidly with J_1 and J_2, a first approximation may be obtained by assuming that Θ is independent of position. In this case (8.15) yields

$$M = \tfrac{1}{4}\pi\psi\Theta(b^4 - a^4), \qquad N = \tfrac{3}{2}\pi e\Theta(b^2 - a^2) \tag{8.17}$$

and J_1 and J_2 can be evaluated at $r = b$. By choosing suitable values of e and ψ the material is subjected to different deformations in which J_1 remains constant and J_2 varies, or vice versa. In this manner the dependence of Θ on J_1, J_2 and t can be determined.

Using their experimental data for a vulcanized synthetic rubber and four filled PVC compounds, Bergen *et al.* (1960) found that Θ was essentially independent of J_2. If this is taken as an assumption, the dependence of Θ on J_1 and t can be determined from simple-extension data using the second of (8.17) and noting that $J_1 = 3e^2$ for this case. These measured values can then be used with (8.17) and the first of (8.16) to predict the behaviour in combined extension torsion tests. Bergen *et al.* (1960) found satisfactory agreement between theoretical predictions and experimental data; this implies that the assumptions made in performing the analysis are adequate for the materials and conditions considered. In addition the data showed that the dependence of Θ on its arguments is of the form

$$\Theta = \phi(t)\,\theta(J_1, J_2) \tag{8.18}$$

This type of separable dependence is considered in greater detail in Chapter 9.

C. Finite Deformations

When deformations are large the material properties are characterized through the functions α_i appearing in (8.7). For an incompressible material, and by analogy with (4.49) and (4.50) for an elastic material, this constitutive relation may be replaced by

$$\boldsymbol{\sigma}(t) = -p\mathbf{I} + \alpha_1\mathbf{B} + \alpha_2\mathbf{B}^2 \tag{8.19}$$

where

$$\alpha_1 = 2\left[\frac{\partial W}{\partial I_1} + I_1\frac{\partial W}{\partial I_2}\right], \qquad \alpha_2 = -2\frac{\partial W}{\partial I_2} \tag{8.20}$$

and $W = W(I_1, I_2, t)$. In the spirit of the discussion at the end of Chapter 4, Section 4, $\partial W/\partial I_1$ and $\partial W/\partial I_2$ may be regarded as independent functions replacing α_1 and α_2. The function W plays a similar role to the strain energy function of elasticity theory, but has an additional dependence here on time.

An experimental study of an unfilled styrene-butadiene vulcanizate in uniaxial and biaxial stress-relaxation has been made by Smith and Frederick (1968). Biaxial tests took the form of extension of a tube in which the internal pressure was adjusted to maintain a constant external diameter. Thus the longitudinal extension ratio is known, the circumferential value is unity (strictly only at the external surface) and the radial value can be calculated from the assumption of incompressibility. Experimental data for the axial stress σ_z and circumferential stress σ_θ were consistent with the forms

$$\sigma_z(\lambda, t, T) = 4G(t, T)\,\Gamma_z(\lambda)$$

$$\sigma_\theta(\lambda, t, T) = 4G(t, T)\,\Gamma_\theta(\lambda) \tag{8.21}$$

where Γ_z and Γ_θ depend only on the axial extension ratio λ, and $G(t,T)$ is the temperature-dependent stress-relaxation function for shear which was assumed to satisfy time-temperature equivalence. Experimental data were obtained for temperatures in the range −40°C to 20°C and extension ratios $\lambda < 2.5$. Separable dependence of the form (8.21) has been found for other rubber vulcanizates in uniaxial extension (Smith, 1962; Halpin, 1965). Whereas Smith and Frederick's (1968) data are modelled by (8.21) within the range of times considered, it is noted that full recovery did not occur during their tests, and (8.21) may not be valid at longer times.

Since $\mathbf{B}$ does not vary for times $t > 0$, (8.19), (8.20) and (8.21) require that $G^{-1}\partial W/\partial I_1$ and $G^{-1}\partial W/\partial I_2$ should be independent of time and temperature, and it remains to determine the dependence of these functions on I_1 and I_2. Data indicated that for $\lambda < 2$ the former is essentially constant, but decreases with deformation for $\lambda > 2$: the second function was found to decrease with λ over the whole range. An adequate model was obtained by assuming that $\partial W/\partial I_1$ is independent of I_2 and $\partial W/\partial I_2$ is independent of I_1; this assumption has been used by other workers (Rivlin and Saunders, 1951; Williams and Major, 1967). Smith and Frederick (1968) used this model and their biaxial data to predict uniaxial stress-relaxation response, and found reasonable agreement between measurements and predictions. For extension ratios $\lambda < 2$ they found that

$$W = G(t, T)\,[a(I_1 - 3) + b\ln(I_2/3)] \tag{8.22}$$

provided an adequate model, where a and b are constants.

For other materials and loading systems the isothermal "strain energy" function $W(I_1, I_2, I_3, t)$ may not have a simple form in which time is a separable effect. In this case the methods employed for finite elastic behaviour must be applied to isochronous data to determine the dependence on the invariants. The derivation of a relatively simple mathematical form for W is then essentially an exercise in surface fitting.

2. SINGLE-STEP CREEP UNDER COMBINED LOAD

Consider the strain response to a loading of arbitrary three-dimensional form which is applied at some time denoted by $t = 0$ and is maintained constant for $t > 0$. For simplicity of analysis, it is assumed that the loading is such that the stress remains constant in a coordinate system which rotates with the rigid-body-rotation component of the resultant deformation. Thus

$$\mathbf{\Sigma}(t) = \mathbf{\Sigma}_0 H(t) \tag{8.23}$$

where $\mathbf{\Sigma}_0$ is independent of t. When this expression is substituted into (4.24) it is evident that

$$\mathbf{E}(t) = \mathbf{f}\{\mathbf{\Sigma}_0, t\} \tag{8.24}$$

where $\mathbf{f}$ is a matrix function. For an isotropic material equation (8.24) has the explicit form

$$\mathbf{E}(t) = \alpha_0 \mathbf{I} + \alpha_1 \mathbf{\Sigma}_0 + \alpha_2 \mathbf{\Sigma}_0^2 \tag{8.25}$$

where α_0, α_1 and α_2 are functions of t and polynomials in the invariants $\operatorname{tr}\mathbf{\Sigma}_0$, $\operatorname{tr}\mathbf{\Sigma}_0^2$ and $\operatorname{tr}\mathbf{\Sigma}_0^3$. In particular, the functions appearing in a third-order theory are

$$\alpha_0 = a_1 \operatorname{tr}\mathbf{\Sigma}_0 + a_2(\operatorname{tr}\mathbf{\Sigma}_0)^2 + a_3 \operatorname{tr}\mathbf{\Sigma}_0^2 + a_4(\operatorname{tr}\mathbf{\Sigma}_0)^3 + a_5 \operatorname{tr}\mathbf{\Sigma}_0^3 + a_6(\operatorname{tr}\mathbf{\Sigma}_0)(\operatorname{tr}\mathbf{\Sigma}_0^2)$$

$$\alpha_1 = b_1 + b_2 \operatorname{tr}\mathbf{\Sigma}_0 + b_3(\operatorname{tr}\mathbf{\Sigma}_0)^2 + b_4 \operatorname{tr}\mathbf{\Sigma}_0^2$$

$$\alpha_2 = c_1 + c_2 \operatorname{tr}\mathbf{\Sigma}_0 \tag{8.26}$$

where a_i, b_i and c_i are functions of t only. These functions can be identified with the creep functions ψ_α of the multiple-integral theory by substituting (8.23) into (4.28) with the definitions (4.27).

For two-dimensional states of stress (8.25) is replaced by

$$\bar{\mathbf{E}}(t) = \beta_0 \bar{\mathbf{I}} + \beta_1 \bar{\mathbf{\Sigma}}_0 \tag{8.27}$$

(see Rivlin, 1955) where the bar denotes a 2×2 matrix and β_0 and β_1 are functions of t and of the invariants $\operatorname{tr}\bar{\mathbf{\Sigma}}_0$ and $\operatorname{tr}\bar{\mathbf{\Sigma}}_0^2$. In a third-order theory, β_0 and β_1 are given by

$$\beta_0 = \pi_1 \operatorname{tr}\bar{\mathbf{\Sigma}}_0 + \pi_2 \operatorname{tr}\bar{\mathbf{\Sigma}}_0^2 + \pi_3(\operatorname{tr}\bar{\mathbf{\Sigma}}_0)^2 + \pi_4(\operatorname{tr}\bar{\mathbf{\Sigma}}_0)(\operatorname{tr}\bar{\mathbf{\Sigma}}_0^2) + \pi_5(\operatorname{tr}\bar{\mathbf{\Sigma}}_0)^3$$

$$\beta_1 = \nu_0 + \nu_1 \operatorname{tr}\bar{\mathbf{\Sigma}}_0 + \nu_2 \operatorname{tr}\bar{\mathbf{\Sigma}}_0^2 + \nu_3(\operatorname{tr}\bar{\mathbf{\Sigma}}_0)^2 \tag{8.28}$$

where π_i and ν_i are functions of t only. When deformations and rotations are small $\mathbf{E}$ and $\mathbf{\Sigma}_0$ may be replaced by $\boldsymbol{\epsilon}$ and $\boldsymbol{\sigma}_0$, provided that the effects of non-linearity are physical rather than geometric in nature.

Onaran and Findley (1965) showed that the material functions π_i and ν_i can be determined in tests of combined extension and torsion. Writing the stress matrix in the form

$$\bar{\mathbf{\Sigma}}_0 \simeq \bar{\boldsymbol{\sigma}}_0 = \begin{Vmatrix} \sigma & T \\ T & 0 \end{Vmatrix} \tag{8.29}$$

it follows from (8.27) and (8.28) that

$$\begin{aligned} \epsilon_{11} &= \sigma(\pi_1 + \nu_0) + \sigma^2(\pi_2 + \pi_3 + \nu_1) + \sigma^3(\pi_4 + \pi_5 + \nu_2 + \nu_3) \\ &\qquad + 2\sigma T^2(\pi_4 + \nu_2) + 2T^2\pi_2 \\ \epsilon_{12} &= T\nu_0 + 2T^3\nu_2 + \sigma T\nu_1 + T\sigma^2(\nu_2 + \nu_3). \end{aligned} \tag{8.30}$$

It is readily seen that the results of three tension tests are sufficient to determine the combinations of material functions which form the coefficients of σ, σ^2 and σ^3 in the expression for ϵ_{11}. In a similar manner two torsion tests are required to determine π_2, ν_0 and ν_2, and two combined tension-torsion tests then give ν_1, ν_3 and π_4. Values of π_1, π_3 and π_5 then follow from the combinations determined in the tension tests. Onaran and Findley (1965) performed these experiments for an unplasticized PVC copolymer and fitted simple empirical expressions to their data [see equation (9.4)].

3. SHORT-TIME APPROXIMATION

A. Three-Dimensional Deformation of an Incompressible Material

Huang and Lee (1966) presented a theory for incompressible materials which is valid in a short time interval following the beginning of a strain programme. The starting point for the theory is equation (6.6) for an incompressible material: a similar theory could be derived for a compressible material by using (4.28), but the derivation is rather tedious. From the point of view of material characterization, the main attraction of the short-time approximation is that it requires only material functions which depend on a single argument.

The first term in (6.6) can be rewritten in the form

$$\int_{-\infty}^{t} \psi_2(t-\tau)\,\dot{\mathbf{E}}(\tau)\,d\tau = \psi_2(0)\,\mathbf{E}(t) - \int_{0}^{t} \frac{\partial}{\partial\tau}\psi_2(t-\tau)\,\mathbf{E}(\tau)\,d\tau \tag{8.31}$$

where it has been assumed that the material is unstrained prior to $t = 0$. When use is made of the symmetry of ψ_6, the second term in (6.6) yields

$$\iint_{-\infty}^{t} \psi_6(t-\tau_1, t-\tau_2)\,\dot{\mathbf{E}}(\tau_1)\,\dot{\mathbf{E}}(\tau_2)\,d\tau_{(2)} =$$

$$\psi_6(0,0)\,\mathbf{E}^2(t) - \int_0^t \frac{\partial}{\partial\tau}\psi_6(t-\tau,0)\,[\mathbf{E}(\tau)\,\mathbf{E}(t) + \mathbf{E}(t)\,\mathbf{E}(\tau)]\,d\tau + \mathbf{R}_2(t) \qquad (8.32)$$

where

$$\mathbf{R}_2(t) = \iint_0^t \frac{\partial^2}{\partial\tau_1\,\partial\tau_2}\psi_6(t-\tau_1, t-\tau_2)\,\mathbf{E}(\tau_1)\,\mathbf{E}(\tau_2)\,d\tau_{(2)}. \qquad (8.33)$$

It can be shown (Huang and Lee, 1966) that for sufficiently small values of t, R_2 can be neglected in comparison with the other contributions in (8.32). The third-order terms in (6.6) can be rewritten in a similar manner and, for sufficiently short times, it follows that the constitutive relation can be written

$$\begin{aligned}\mathbf{R}^T\mathbf{sR} + p\mathbf{I} = a\mathbf{E} + b\mathbf{E}^2 + c\mathbf{E}\,\mathrm{tr}\,\mathbf{E}^2 + d\mathbf{E}^3 \\ - \int_0^t \{\gamma_1\,\mathbf{E}(\tau) + \gamma_2[\mathbf{E}(\tau)\,\mathbf{E} + \mathbf{E}\mathbf{E}(\tau)] + \gamma_3\,\mathbf{E}(\tau)\,\mathrm{tr}\,\mathbf{E}^2 \\ + \gamma_4\,\mathbf{E}\,\mathrm{tr}\,[\mathbf{E}\mathbf{E}(\tau)] + \gamma_5[\mathbf{E}(\tau)\,\mathbf{E}^2 + \mathbf{E}^2\,\mathbf{E}(\tau)] + \gamma_6\,\mathbf{E}\mathbf{E}(\tau)\,\mathbf{E}\}\,d\tau\end{aligned} \qquad (8.34)$$

where $\mathbf{E} \equiv \mathbf{E}(t)$. The material constants a, b, c and d are given by

$$a = \psi_2(0), \qquad b = \psi_6(0,0), \qquad c = \psi_{10}(0,0,0), \qquad d = \psi_{12}(0,0,0) \qquad (8.35)$$

and the γ_i are functions of $t - \tau$ defined by

$$\gamma_1(t-\tau) = \frac{\partial}{\partial\tau}\psi_2(t-\tau) \qquad \gamma_2(t-\tau) = \frac{\partial}{\partial\tau}\psi_6(t-\tau,0)$$

$$\gamma_3(t-\tau) = \frac{\partial}{\partial\tau}\psi_{10}(t-\tau,0,0) \qquad \gamma_4(t-\tau) = \frac{\partial}{\partial\tau}\psi_{10}(0,t-\tau,0)$$

$$\gamma_5(t-\tau) = \frac{\partial}{\partial\tau}\psi_{12}(t-\tau,0,0) \qquad \gamma_6(t-\tau) = \frac{\partial}{\partial\tau}\psi_{12}(0,t-\tau,0). \qquad (8.36)$$

The first four terms on the right of (8.34) represent the third-order approximation to the instantaneous elastic response due to $\mathbf{E}(t)$. The integral term is the contribution due to the deformation history.

B. Determination of Material Functions

The material parameters appearing in (8.34) are special values of those appearing in (6.6); the required values are indicated in (8.35) and (8.36). In principle these values can be obtained from an experimental programme which is a subset of the tests required for (6.6). A possible procedure is to follow tests 1–6 of Table 3 in Chapter 6 and to measure outputs in tests 1 and 2 at $t = \epsilon$ and in tests 3–6 at $t = k + \epsilon$, where ϵ is a small quantity. The analysis of Section 2C of Chapter 6 can then be followed and, in particular, expressions (6.12), (6.13), (6.17), (6.18) and (6.19) yield

$$\psi_2(\epsilon), \qquad \psi_6(\epsilon, \epsilon), \qquad \psi_{10}(\epsilon, \epsilon, \epsilon), \qquad \psi_{12}(\epsilon, \epsilon, \epsilon) \tag{8.37}$$

$$\psi_6(t, \epsilon), \qquad \psi_{10}(t, \epsilon, \epsilon), \qquad \psi_{10}(\epsilon, t, \epsilon) \tag{8.38}$$

$$\psi_{12}(\epsilon, t, \epsilon) + 2\psi_{12}(t, \epsilon, \epsilon). \tag{8.39}$$

In a similar manner, test 9 of Table 3 in Chapter 6 and expression (6.26) gives

$$2\psi_{12}(t, \epsilon, \epsilon) - \psi_{12}(\epsilon, t, \epsilon). \tag{8.40}$$

In the limit as $\epsilon \to 0$ the values (8.37)–(8.40) yield values for all of the material functions required in (8.35) and (8.36). This completes the determination of material parameters for the short-time approximation and, by reference to Table 3 in Chapter 6, the total number of tests is $5N + 2$. In the present case, however, the motivation for the choice of N is different from that considered previously since t, and therefore k, need take only small values and a smaller value of N may be adequate to cover the time interval of interest. If it is sufficient to represent, for example, $\psi_6(t, 0)$ by a low-order polynomial

$$\psi_6(t, 0) = b_0 + \sum_{n=1}^{M} b_n t^n \tag{8.41}$$

for some M, then N can be taken equal to M. In the simplest case where $M = N = 1$ only seven experiments are required and the parameters a, b, c, d and γ_i are all constants. Equation (8.34) is then a matrix polynomial in $\mathbf{E}$ and linear in $\mathbf{A}$ where

$$\mathbf{A}(t) = \int_0^t \mathbf{E}(\tau)\, d\tau. \tag{8.42}$$

A systematic theory of this type for one-dimensional behaviour in which the kernel functions are expanded as power series in their time-like arguments has been given by Lubliner (1967).

The experimental programme proposed in this sub-section may not be sufficiently accurate to use in practice, since it relies on measurements taken a very short time ϵ after a deformation change. and inertia effects may then make data unreliable. In this case the theory and the experimental programme have greater validity for continuous strain histories, and the power series development of Lubliner (1967) takes a simpler form.

4. SMALL DEFORMATIONS SUPERPOSED ON FINITE DEFORMATIONS

When the deformation of a material varies only slightly during the time scale of interest, separation of the deformation into a time-independent finite component and a time-dependent small component is not unique. Two particular cases are considered here. The first is relevant to the situation in which the material has been subjected to a finite deformation for a time longer than the characteristic memory time of the material. In this case the deformation at time t, the instant at which stress is to be evaluated, can be taken as the finite constant deformation and this can be assumed to have been applied for all time. In the second case, the material is undeformed up to time $\tau = 0$ and the instantaneous elastic response at $\tau = 0$ is chosen as the finite deformation. For times $\tau > 0$ the deformation varies only slightly from this value.

A. Variation About Current Deformation

When the deformation at time t is chosen as the finite component, the deformation at time τ can be written in the form

$$x_i(X_j, \tau) = x_i(X_j, t) + u_i(X_j, \tau) \tag{8.43}$$

where u_i are small displacements with respect to which a linearized theory is to be derived. From (8.43) it follows that $\mathbf{C}(\tau)$ can be expressed as

$$\mathbf{C}(\tau) = \mathbf{C}(t) + \mathbf{K}(\tau) \tag{8.44}$$

where

$$\mathbf{K}(\tau) = \frac{\partial u_k}{\partial X_i}\frac{\partial x_k}{\partial X_j} + \frac{\partial x_k}{\partial X_i}\frac{\partial u_k}{\partial X_j} \tag{8.45}$$

and $x_k \equiv x_k(t)$. Substitution of this expression into (4.34) gives

$$\boldsymbol{\sigma}(t) = \mathbf{F}(t) \underset{\tau=-\infty}{\overset{t}{\mathscr{F}}} \{\mathbf{C}(t) + \mathbf{K}(\tau)\}\, \mathbf{F}^T(t) \tag{8.46}$$

and linearization with respect to **K** yields

$$\boldsymbol{\sigma}(t) = \mathbf{F}(t) \underset{\tau=-\infty}{\overset{t}{\mathscr{L}}} \{\mathbf{K}(\tau), \mathbf{C}(t)\} \mathbf{F}^T(t) \tag{8.47}$$

where $\mathscr{L}$ is a linear matrix functional of $\mathbf{K}(\tau)$, but is a nonlinear function of $\mathbf{C}(t)$.

An alternative form to (8.47) can be readily derived in which the relative strain measure $\mathbf{C}_t(\tau)$ replaces (8.44). These two theories are similar, respectively, to a relation derived by Pipkin and Rivlin (1961) and to the theory of finite linear viscoelasticity derived by Coleman and Noll (1961). In both cases the constitutive equation can be written in an explicit form in which only single integrals occur and material functions depend on a single time-like variable. However, these stress-relaxation functions depend in a general manner on the finite deformation component and, as was seen in Chapter 7, Section 2, a number of further approximations have to be made in order to achieve a tractable theory.

B. Small Deformation Superposed on Instantaneous Deformation

In this case the material is supposed to be in its reference configuration X_i for all times $\tau < 0$, and is given a finite deformation at $\tau = 0$. The deformation is assumed to vary only slightly for $\tau \geqslant 0$, so that

$$x_i(X_j, \tau) = X_i, \qquad \tau < 0$$

$$x_i(X_j, \tau) = x_i(X_j, 0) + u_i(X_j, \tau), \qquad \tau \geqslant 0 \tag{8.48}$$

where u_i are small displacements with respect to which a linearized theory is to be derived. From (8.48) it follows that

$$\mathbf{R}(t) = \mathbf{R}(0) + \mathbf{R}_1(t), \qquad \mathbf{F}(t) = \mathbf{F}(0) + \mathbf{F}_1(t), \qquad t > 0$$

$$\mathbf{E}(\tau) = \mathbf{E}(0)\, H(\tau) + \mathbf{E}_1(\tau)$$

$$2\mathbf{E}_1(\tau) = \mathbf{F}^T(0)\, \mathbf{F}_1(\tau) + \mathbf{F}_1^T(\tau)\, \mathbf{F}(0)$$

$$\mathbf{F}(0) = \left\| \frac{\partial x_i(X_k, 0)}{\partial X_j} \right\|, \qquad \mathbf{F}_1(\tau) = \left\| \frac{\partial u_i(X_k, \tau)}{\partial X_j} \right\| \tag{8.49}$$

where $\mathbf{R}_1$, $\mathbf{F}_1$ and $\mathbf{E}_1$ are small components.

Within this notation (4.16) can be written in the form

$$\boldsymbol{\sigma}(t) = [\mathbf{R}(0) + \mathbf{R}_1(t)] \underset{\tau=-\infty}{\overset{t}{\mathscr{H}}} \{\mathbf{E}(0)\, H(\tau) + \mathbf{E}_1(\tau)\} [\mathbf{R}^T(0) + \mathbf{R}_1^T(t)] \tag{8.50}$$

where the functional can be expressed as a linear functional in $\mathbf{E}_1(\tau)$ in which the material functions depend on $\mathbf{E}(0)$, t and τ. By assuming an explicit form, the dependence of the material functions on the finite component $\mathbf{E}(0)$ can be avoided. This is most easily done when the material is incompressible and a third-order theory in $\mathbf{E}(0)$ is considered. In that case the functional $\mathscr{H}$ can be written in the form (6.6) with

$$\mathbf{M}(\tau) = \mathbf{E}(0)\,\delta(\tau) + \dot{\mathbf{E}}_1(\tau). \tag{8.51}$$

It follows that $\mathscr{H}$ can be expressed as the sum of a large component $\mathscr{H}_0$ and a small component $\mathscr{H}_1$, where

$$\mathscr{H}_0 = \mathbf{E}(0)\,\psi_2(t) + \mathbf{E}^2(0)\,\psi_6(t,t) + \mathbf{E}(0)\operatorname{tr}\mathbf{E}^2(0)\,\psi_{10}(t,t,t) + \mathbf{E}^3(0)\,\psi_{12}(t,t,t) \tag{8.52}$$

$$\begin{aligned}\mathscr{H}_1 = {} & \int_{-\infty}^{t} \psi_2 \dot{\mathbf{E}}_1(\tau)\,d\tau + \int_{-\infty}^{t} \psi_6(t, t-\tau)\,[\mathbf{E}(0)\,\dot{\mathbf{E}}_1(\tau) + \dot{\mathbf{E}}_1(\tau)\,\mathbf{E}(0)]\,d\tau \\ & + \int_{-\infty}^{t} \psi_{10}(t,t,t-\tau)\operatorname{tr}\mathbf{E}^2(0)\,\dot{\mathbf{E}}_1(\tau)\,d\tau + 2\int_{-\infty}^{t} \psi_{10}(t, t-\tau, t) \\ & \times \operatorname{tr}[\mathbf{E}(0)\,\dot{\mathbf{E}}_1(\tau)]\,\mathbf{E}(0)\,d\tau + \int_{-\infty}^{t} \psi_{12}(t, t-\tau, t)\,\mathbf{E}(0)\,\dot{\mathbf{E}}_1(\tau)\,\mathbf{E}(0)\,d\tau \\ & + 2\int_{-\infty}^{t} \psi_{12}(t,t,t-\tau)\,\mathbf{E}^2(0)\,\dot{\mathbf{E}}_1(\tau)\,d\tau.\end{aligned} \tag{8.53}$$

Finally, the constitutive relation becomes

$$\begin{aligned}\boldsymbol{\sigma}(t) = {} & -p\mathbf{I} + \mathbf{R}(0)\mathscr{H}_0\,\mathbf{R}^T(0) + \mathbf{R}_1(t)\mathscr{H}_0\,\mathbf{R}^T(0) + \mathbf{R}(0)\mathscr{H}_1\,\mathbf{R}^T(0) \\ & + \mathbf{R}(0)\mathscr{H}_0\,\mathbf{R}_1^T(t).\end{aligned} \tag{8.54}$$

Since the theory considered above is a special form of the theory discussed in Chapter 6, Section 2, the experimental programme for determining the material functions appearing in (8.52) and (8.53) is a subset of those indicated in Table 3 of Chapter 6. Experiments 1–6 and 9 are sufficient to determine ψ_2 and ψ_6 completely and ψ_{10} and ψ_{12} when their arguments are not all distinct. However, some reduction in the required number of experiments is achieved here since ψ_{10} and ψ_{12} are required only when the equal arguments are greater than the third argument. Thus, $\psi_{10}(t,t,t-\tau)$ is required, but $\psi_{10}(t-\tau, t-\tau, t)$ is not needed. This simplification of the experimental programme is achieved by making B a small quantity in (6.14) and in experiment 9. It follows that terms in B^2 and Y^2 can be omitted from (6.16) and (6.25) and only two tests of type (6.14) need be performed for each k. Thus, the total number of experiments in this case is $3N + 2$.

An alternative constitutive relation has been considered by Appleby and Lee (1966) in which **R** is replaced by **F** in (8.50) *et seq*. This form may be preferable for the solution of boundary value problems, since **F** is more easily expressed in terms of the deformation than is **R**. However, the analysis of experimental data becomes more complex since even in simple extension the functional $\mathscr{H}$ in the equation replacing (8.50) is pre- and post-multiplied by a non-unit matrix. Appleby and Lee (1966) claim that the material functions in (8.52) and (8.53) can be determined entirely in simple extension tests; this is not the case since for such deformations ψ_{12} always appears in the combination (6.19). A test like experiment 9 of Table 3 in Chapter 6 is required to obtain the separate values of $\psi_{12}(t, t-\tau, t)$ and $\psi_{12}(t, t, t-\tau)$.

5. ALMOST-ELASTIC MATERIALS

In the previous sub-section situations were considered in which the deformation varies only slightly after the instantaneous elastic response at $\tau = 0$. Since, in general, the stress required to produce a constant deformation is not approximately time-independent, the stress histories resulting from such deformations do not consist of a finite time-independent component and a small time-dependent component. A dual theory could be readily derived from equation (4.24), in which stress $\mathbf{\Sigma}$ varies only slightly after the instantaneously applied value at $\tau = 0$ but, in general, the dual theories would refer to mutually exclusive sets of circumstances. For this reason, interrelations between the material functions of the two theories cannot be established. The exception to this statement is the special case of materials whose behaviour is almost elastic; in this case an almost constant deformation is supported by an almost constant stress. A theory for almost-elastic materials has been proposed by Lockett (1969).

An almost-elastic material is such that the stress depends strongly on the current strain, but only slightly on strain history. Thus (4.16) is replaced by

$$\mathbf{\Sigma}(t) = \mathbf{f}\{\mathbf{E}(t)\} + \eta \underset{\tau=-\infty}{\overset{t}{\mathscr{F}}} \{\mathbf{E}(\tau)\} \tag{8.55}$$

where **f** is a matrix function and η is a small parameter. In addition to differences in choice of stress and strain measure, (8.55) differs from (8.47) in that $\mathscr{F}$ represents a nonlinear functional, whereas $\mathscr{L}$ is a linear functional. When $\eta = 0$, (8.55) represents an elastic material and, for any boundary value problem $\mathbf{E}_0(t)$ is used to denote the corresponding elastic solution for the strain. For the same boundary value problem, the strain in the viscoelastic material denoted by (8.55) is slightly different from $\mathbf{E}_0(\tau)$ and may be written

$$\mathbf{E}(\tau) = \mathbf{E}_0(\tau) + \eta \mathbf{E}_1(\tau). \tag{8.56}$$

Substitution of (8.56) in (8.55) yields, to first order in η,

$$\mathbf{\Sigma}(t) = \mathbf{f}\{\mathbf{E}_0(t)\} + \eta \mathbf{E}_1(t)\, \partial \mathbf{f}\{\mathbf{E}_0(t)\}/\partial \mathbf{E}_0(t) + \eta \underset{\tau=-\infty}{\overset{t}{\mathscr{F}}} \{\mathbf{E}_0(\tau)\}. \tag{8.57}$$

The first term in (8.57) represents the elastic solution corresponding to $\mathbf{E}_0$: the other two terms are the perturbation due to the slight viscoelasticity. The first of these terms is proportional to $\mathbf{E}_1(t)$ with a "modulus" dependent on the elastic strain $\mathbf{E}_0(t)$. The last term in (8.57) has the effect of a body force depending on $\mathbf{E}_0(t)$. It is noted that the functional $\mathscr{F}$ has to be evaluated at $\mathbf{E} = \mathbf{E}_0$; this simplifies stress analysis, but does not in general reduce the experimental problem.

This may be achieved by considering situations where boundary conditions are applied at some instant $\tau = 0$ and are held constant thereafter. Although neither the stress nor strain field is time-independent for a viscoelastic material; in this case it follows that

$$\mathbf{E}_0(\tau) = \mathbf{E}_0\, H(\tau) \tag{8.58}$$

provided that inertia effects are negligible, and the functional in (8.57) then becomes

$$\underset{\tau=-\infty}{\overset{t}{\mathscr{F}}} \{\mathbf{E}_0(\tau)\} = \mathbf{g}\{\mathbf{E}_0, t\}. \tag{8.59}$$

For a third-order theory in $\mathbf{E}_0$ (8.59) involves twelve functions of t which can be identified with the functions ψ_α of (4.28) evaluated with all arguments equal to t. For an incompressible material only the functions $\psi_2(t)$, $\psi_6(t,t)$, $\psi_{10}(t,t,t)$ and $\psi_{12}(t,t,t)$ are involved and these functions (of a single variable) are readily determined from experiments 1 and 2 of Table 3, Chapter 6. The dual creep formulation, and interrelations between creep and stress-relaxation functions, have been considered by Lockett (1969).

Chapter 9 Semi-Empirical Approach to Material Characterization

Two approaches to material characterization have been made in the literature: the empirical approach, in which experiments are motivated by qualitative reasoning and experimental data are modelled by curve fitting to simple mathematical forms, and the rational approach, in which theories are constructed around a set of physical axioms and thereafter the development satisfies a high degree of rigour. The first method can produce results for special materials and loading situations in a relatively quick and simple manner. However, such models are generally crude and cannot be applied to new loading situations with any degree of confidence. For less sophisticated applications this may be adequate and, until the last few years, the empirical approach could be supported by the philosophy that something is better than nothing. However, as polymers are being considered for more sophisticated applications it has become obvious that greater sophistication is required in the method of material characterization. This has been the motivation behind the rational theories proposed in the last fifteen years, but it should be obvious from the previous chapters of this book that this approach has not reached a stage of development where it can replace the empirical method at a practical level, except in certain special circumstances.

It would appear that the approach most likely to succeed at the present time is one in which the empirical and rational approaches are combined. Since the rational approach would be preferable if it was capable of application, it should be the starting point with empiricism being introduced into the development at as late a stage as possible. The present chapter describes work which can be considered in this context, and the entire chapter is intended to give an indication of the direction from which progress is likely to come. However, since the characterization of nonlinear viscoelastic behaviour is still an active research area, it is not possible to give more than a general impression from which the reader can draw his own conclusions.

1. EMPIRICAL REPRESENTATION

It is useful to begin with a brief discussion of the type of relation used in empirical theories. No attempt is made to give a survey of such representations here, but a few particular examples are discussed which can be used later to motivate semi-empirical methods of characterization.

Many experiments have been reported in the literature involving creep in various modes of deformation and at stress levels exceeding the linear limit. For these single-step creep tests the strain is a function of time and of the stress level, and attempts have been made to represent the creep data by a relation of the form

$$\epsilon_c(\sigma_0, t) = a(\sigma_0) + b(\sigma_0) f(t). \tag{9.1}$$

The first term represents an instantaneous elastic component; the second term represents a viscoelastic contribution in which time and stress dependence are separable.

Although no particular problems would attach to assuming general dependences of the functions a, b and f on their arguments, empirical representations have usually sought to model these functions through simple analytical expressions involving only a few constant parameters. It has been found that the time-dependence can often be modelled adequately via a simple power law:

$$f(t) = t^n \tag{9.2}$$

where n is a constant; this is simpler than, and tends to be as accurate as, functions containing several exponential terms representing a number of retardation times. Data satisfying (9.1) has been presented by Findley and Khosla (1955), O'Connor and Findley (1962), Findley (1960), Onaran and Findley (1963, 1965), Turner (1966) and others.

Two types of stress dependence are worthy of note, one involving hyperbolic sines and the other a polynomial representation. It has been shown by Findley (1960) and others that creep of many plastics at constant stress σ may be represented adequately by

$$\epsilon_c = \epsilon_a \sinh(\sigma/\sigma_a) + \epsilon_b t^n \sinh(\sigma/\sigma_b) \tag{9.3}$$

where ϵ_a, ϵ_b, σ_a and σ_b are constants. In some cases (see, for example, Onaran and Findley, 1965) σ_a is sufficiently large that $\sinh(\sigma/\sigma_a) \simeq \sigma/\sigma_a$, and the hyperbolic sine is required only in the viscoelastic term. Onaran and Findley (1965) found that their data for an unplasticized PVC could be modelled in pure tension or in pure shear by a relation of this form, the time exponent n being the same in each case.

Since the two forms of (9.3) for tension and shear are purely empirical relations for these separate types of deformation, they cannot be expected to predict adequately the response in more general deformations. In a combined tension/torsion test they provide no prediction for the interaction effect between the two types of deformation. Onaran and Findley (1965) were able to make predictions for this synergistic effect by using the rational theory of Chapter 8,

Section 2, combined with some of the empirical results mentioned above. Thus (9.1) and (9.2), in which $a(\sigma_0)$ and $b(\sigma_0)$ are expressed as polynomials, suggest that the π_i and ν_i in (8.28) should take the form

$$\pi_i = \pi_i^{(0)} + \pi_i^{(1)} t^n, \text{ etc.} \tag{9.4}$$

where $\pi_i^{(0)}$ and $\pi_i^{(1)}$ are constants. Experimental strain data can now be expressed as

$$\epsilon_c(\sigma_0, t) = \epsilon_0(\sigma_0) + \epsilon_1(\sigma_0) t^n \tag{9.5}$$

where, for each σ_0, ϵ_0 is chosen to make the plot of $\epsilon - \epsilon_0$ versus $\log t$ a straight line and ϵ_1 and n are determined from the intercept and slope. In this manner the experiments discussed in Chapter 8, Section 2 can be used to determine $\pi_i^{(0)}$, $\pi_i^{(1)}$, $\nu_i^{(0)}$ and $\nu_i^{(1)}$. These data can be used to predict behaviour in other combined tension-torsion experiments; Onaran and Findley (1965) found that the predictions agreed satisfactorily with their experimental data.

2. LEADERMAN'S CONSTITUTIVE RELATION

In the case of linear infinitesimal behaviour it was seen that the one-dimensional constitutive relation can be expressed in the form (2.9):

$$\epsilon(t) = \int_{-\infty}^{t} \psi(t - \tau)\, \dot{\sigma}(\tau)\, d\tau. \tag{9.6}$$

For nonlinear behaviour this equation may be replaced by the Green-Rivlin representation (5.1) in which $P = \sigma$ and Q may represent ϵ or a nonlinear measure such as E. If $Q = \epsilon$, nonlinear behaviour cannot be modelled by the first term on the right-hand side of (5.1): equation (5.1) is a generalization of (9.6) in which the right-hand side of the latter equation is modified. Leaderman (1962) proposed that an alternative generalization of (9.6) would be to modify the left-hand side of that equation by replacing ϵ with a suitable nonlinear strain measure. In this context suitability would require that the new strain measure could be adequately represented by a linear functional of the form (9.6). The results of Chapter 5 indicate that, although E is a candidate for this nonlinear strain measure, it does not lead to a linear relation. Leaderman's (1962) postulate can be expressed as

$$f[\epsilon(t)] = \rho[\lambda(t)] = \int_{-\infty}^{t} \psi(t - \tau)\, \dot{\sigma}(\tau)\, d\tau \tag{9.7}$$

where, for extensional behaviour, an alternative description in terms of the extension ratio $\lambda(t) = 1 + \epsilon(t)$ can be considered. The form of function ρ chosen by Leaderman (1962) is

$$\rho(\lambda) = (\lambda^2 - \lambda^{-1})(1 + k\lambda^{-1})/3(1 + k) \tag{9.8}$$

where k is a constant and, in agreement with previous notation, σ denotes the true stress in the deformed material. The denominator in (9.8) is chosen so that for small strains $\rho(\lambda) \to \epsilon$ and thus the creep function ψ in (9.7) can be identified with that in (9.6).

The particular form of strain measure (9.8) is motivated by results of Rivlin and Saunders (1951) for the nonlinear mechanical behaviour of incompressible elastic materials; this has been discussed in Chapter 4. This form is also familiar from equations (7.22) and (7.23) for the BKZ and Lianis theories, where adequate modelling was obtained for some materials with $\alpha \equiv 0$. Indeed the motivations for (9.8) and for the BKZ and Lianis theories depend in a similar manner on the results of nonlinear elasticity.

Using (5.5) for extensional behaviour it is readily seen that $E = \frac{1}{2}(\lambda^2 - 1)$, so that (9.8) is no more a simple function of E than it is of ϵ.

It has been noted that σ in (9.7) and elsewhere denotes true stress; Leaderman (1962) also considered a formulation in terms of the nominal stress σ_n, that is, the stress defined with respect to area in the undeformed material. For this case Leaderman (1962) considered

$$\rho_n(\lambda) \equiv (\lambda - \lambda^{-2})(1 + k_n\lambda^{-1})/3(1 + k_n) = \int_{-\infty}^{t} \psi_n(t - \tau)\,\dot{\sigma}_n(\tau)\,d\tau \tag{9.9}$$

where the suffix n has been used in ρ_n, k_n and ψ_n to identify these quantities with the nominal-stress formulation. For extension of an incompressible material σ and σ_n are related by $\sigma = \lambda\sigma_n$; however, since λ is time dependent in general, (9.9) does not reduce to (9.7) with (9.8), and k_n, ψ_n will be different from k, ψ.

A further possibility which Leaderman (1962) considered for generalizing (9.7) or (9.9) is to replace the stress σ or σ_n by some nonlinear stress measure $F_1(\sigma)$ or $F_2(\sigma_n)$; however, this was found to be unnecessary for the experimental data discussed by Leaderman (1962).

Since (9.7) is a linear functional it can be inverted to give a corresponding stress-relaxation form

$$\sigma(t) = \int_{-\infty}^{t} \phi(t - \tau)\frac{\partial}{\partial\tau}[f\{\epsilon(\tau)\}]\,d\tau \tag{9.10}$$

where, since $f(\epsilon) \to \epsilon$ as $\epsilon \to 0$, ϕ is the stress-relaxation function of classical linear infinitesimal theory. However, it should be noted that ψ and ϕ can be identified with the functions of linear theory only for a material which is modelled *exactly* by (9.7) and (9.10). When data can be modelled only approximately by these relations, this identification may have to be relaxed slightly in order to obtain approximate representation over the nonlinear region. A three-dimensional analogue of (9.10) has been proposed by Wriggles and Williams (1969).

In a test such as creep-and-recovery, linearity of the relation connecting ρ and σ ensures that simple superposition applies with ρ replacing ϵ in the results of linear theory. Thus, using a notation corresponding to that of Section 3B, in Chapter 5, the recovery response is given by

$$\rho_r(\sigma_0, k; t) = \rho_c(\sigma_0, t) - \rho_c(\sigma_0, t - k). \tag{9.11}$$

Since ρ is a nonlinear function of ϵ (or E) this does not imply a similar expression for ϵ. Thus (9.11) is a possible relation for representing nonlinear behaviour observed with respect to the strain measure ϵ.

Experimental creep and recovery data for plasticized PVC have been analysed by Leaderman (1962) using the representations (9.7)–(9.8) and (9.9); data were considered for various stress levels resulting in elongations of up to 20%. For the formulation (9.9) in terms of nominal stress it was found that the data could be represented accurately when $k_n \equiv 0$. However, the formulation (9.7)–(9.8) in terms of true stress required a strain measure in which $k = 0.8$. The two creep functions ψ and ψ_n were found to differ by a small but significant amount.

In a creep-and-recovery test in which, for example, the true stress σ has a piecewise constant time-dependence

$$\sigma(\tau) = \sigma_0[H(\tau) - H(\tau - k)] \tag{9.12}$$

any function $F(\sigma)$, which is such that $F(0) = 0$, becomes

$$F[\sigma(\tau)] = F(\sigma_0)[H(\tau) - H(\tau - k)]. \tag{9.13}$$

It follows that (9.11) continues to hold when stress in the integrands of (9.7) and (9.9) is replaced by a nonlinear stress measure. It has been noted that Leaderman's (1962) experimental data did not require this latter generalization.

The theory and experimental data discussed above indicate that it is possible, at least in restricted circumstances, to model nonlinear behaviour through a *linear* functional connecting nonlinear measures of stress and/or strain. The question naturally arises as to which measures are the most appropriate. As far as stress is concerned, the choice lies between true and nominal stress: nonlinear functions of these quantities may be considered, but this has not been a popular approach (but see Section 7). The question of the "proper" choice of strain measure has received a great deal of discussion in the literature. In general, proposed measures have revolved around use of $\lambda - \lambda^{-2}$ with nominal stress (see, for example, Arai and Suzuki, 1961) or $\lambda^2 - \lambda^{-1}$ with true stress (see Staverman and Schwarzl, 1956); these correspond to $k=0$ or $k_n=0$ in Leaderman's theory, but it is noted that a non-zero value for k was required for his data. All of these suggestions must be regarded as empirical, and no doubt there will be further discussion of what constitutes the most appropriate choice of strain measure. Little consideration has been given to the analogous problem of choice of nonlinear measures for three-dimensional behaviour.

3. RELATION BETWEEN LEADERMAN'S THEORY AND GREEN-RIVLIN FORMULATION

It is evident that the formulation discussed in the previous section is less general than the Green-Rivlin formulation, since the former involves only a linear functional. The relation between the two theories is determined by using (9.8) or (9.9) to express ϵ as a polynomial in ρ or ρ_n:

$$\epsilon = a\rho + b\rho^2 + c\rho^3 + \cdots$$

$$\epsilon = a_n\rho_n + b_n\rho_n^2 + c_n\rho_n^3 + \cdots \tag{9.14}$$

It is readily shown that

$$a = a_n = 1, \qquad b = \frac{k}{1+k}, \qquad c = \frac{2k^2 - 5k - 1}{3(1+k)^2}$$

$$b_n = \frac{1+2k_n}{1+k_n}, \qquad c_n = \frac{2(1+5k_n+7k_n^2)}{3(1+k_n)^2}, \tag{9.15}$$

When (9.7) is substituted into the first of (9.14) there follows an equation of the form

$$\epsilon(t) = \int_{-\infty}^{t} \Phi_1(t-\tau_1)\,\dot{\sigma}(\tau_1)\,d\tau_1 + \iint_{-\infty}^{t} \Phi_2(t-\tau_1, t-\tau_2)\,\dot{\sigma}(\tau_1)\,\dot{\sigma}(\tau_2)\,d\tau_{(2)} + \cdots \tag{9.16}$$

where

$$\Phi_1(s) = \psi(s), \qquad \Phi_2(s_1, s_2) = b\psi(s_1)\,\psi(s_2),$$

$$\Phi_3(s_1, s_2, s_3) = c\psi(s_1)\,\psi(s_2)\,\psi(s_3). \tag{9.17}$$

In a similar manner, when the second of (9.14) is used in conjunction with (9.9) there results an equation like (9.16) in which σ_n replaces σ and the quantities Φ_α are given by (9.17) in which b_n, c_n, ψ_n replace b, c and ψ.

Since (9.16) is the Green-Rivlin constitutive relation, it follows from (9.17) that Leaderman's form is obtained by making the higher order creep functions proportional to products of the first-order function. In particular, Leaderman *et al.* (1963) noted that Leaderman's (1962) data for plasticized PVC, which could be modelled by (9.9) with $k_n = 0$, could also be modelled by a Green-Rivlin formulation (9.16) in which σ is replaced by nominal stress and the Φ's are given by (9.17) with $b_n = 1$, $c_n = \frac{2}{3}$.

The strain measures ρ and ρ_n were chosen so that they reduced to the classical strain measure for small strains; it follows that $a = a_n = 1$. The pair of parameters (b, c) are defined in (9.15) in terms of the single parameter k; similarly (b_n, c_n) are defined in terms of k_n. It is evident that a more general strain measure than that introduced by Leaderman (1962) will lead to an expression (9.14) in which b and c are independent parameters. In this case the quantities b, c and ψ can be treated as unknown material parameters to be determined by experiment. Thus, in a creep test, (9.16) and (9.17) give

$$\epsilon_c(\sigma_0, t) = \sigma_0 \psi(t) + b\sigma_0^2[\psi(t)]^2 + c\sigma_0^3[\psi(t)]^3 + \cdots \tag{9.18}$$

If isochrones of ϵ_c/σ_0 versus σ_0 are plotted, the intercepts give $\psi(t)$; b and c are to be determined from the shape of the isochrones. In this manner Leaderman's (1962) PVC data would have yielded the results $b = 1$, $c = \frac{2}{3}$, thus determining the strain measure.

For the recovery phase of a creep-and-recovery test it is readily shown, with the notation of Section 3B, Chapter 5 that

$$\epsilon_r(\sigma_0, k; t) = \epsilon_c(\sigma_0, t-k) - 2\sigma_0^2[1 + \sigma_0 \Phi_1(t)]\, \Phi_1(t-k)\, [\Phi_1(t-k) - \Phi_1(t)] \tag{9.19}$$

for the PVC creep data. Equation (9.19) allows the model calculated entirely from creep data to be used to compare predictions of recovery behaviour with experimental data. For this purpose isochrones of $(\epsilon_r - \epsilon_c)/\sigma_0^2$ versus σ_0 can be compared with the straight lines predicted by (9.19). Leaderman *et al.* (1963) claim that the agreement is satisfactory; however, it is evident that the agreement is poorer than that obtained in Leaderman's (1962) earlier analysis using (9.9). This is presumably due to the fact that (9.9) and the nominal-stress version of (9.16) and (9.17) are not identical, the latter being a third-order approximation of the former. These comments hint at an argument against the use of polynomial representations in favour of other simple-function relations. This point will be taken up again in Section 6.

4. PRODUCT FORM FOR MATERIAL FUNCTIONS

A. One-Dimensional Relations

The limited success of a Green-Rivlin constitutive relation in which the creep functions are given by (9.17) suggests that material functions in other formulations may be expressible as products of functions of single arguments.

Stafford (1969) suggested that the functions ψ_α in (4.28) may be expressible in the form

$$\psi_\alpha(s_1, s_2, \ldots, s_n) = f_\alpha^{(1)}(s_1)\, f_\alpha^{(2)}(s_2) \ldots f_\alpha^{(n)}(s_n) \tag{9.20}$$

where the symmetries (4.30) imply equalities between some of the $f_i^{(\alpha)}$. In the one-dimensional theory (5.1) the assumption (9.20) and symmetry of Φ_2 and Φ_3 require that

$$\Phi_2(s_1, s_2) = f_2(s_1)\, f_2(s_2), \qquad \Phi_3(s_1, s_2, s_3) = f_3(s_1)\, f_3(s_2)\, f_3(s_3) \tag{9.21}$$

and f_2 and f_3 are known once $\Phi_2(t,t)$ and $\Phi_3(t,t,t)$ are known, since

$$f_2(t) = [\Phi_2(t,t)]^{1/2}, \qquad f_3(t) = [\Phi_3(t,t,t)]^{1/3}. \tag{9.22}$$

Stafford (1969) used values of $\Phi_2(t,t)$ and $\Phi_3(t,t,t)$, determined from the experimental data of Lifshitz and Kolsky (1967) for polyethylene, to calculate the behaviour in a two-step creep test as predicted by (5.1) and (9.21). The predictions for stresses in the nonlinear range, and for relatively long intervals between loading steps, agreed well with experimental data of Lifshitz and Kolsky (1967). For relatively short intervals the predictions were significantly lower than the experimental data. However, the error was time-independent over the time scale considered, and Stafford (1969) concluded that the error was of an elastic nature which did not invalidate the product form for the viscoelastic response. This conclusion is supported by a slightly different form of material function postulated by Findley and Onaran (1968). Further discussion of Stafford's (1969) theory is given in Section 5.

Findley and Onaran (1968) expressed the creep functions Φ_α in the form

$$\Phi_\alpha(s_1, s_2, \ldots, s_\alpha) = \mu_\alpha + f_\alpha(s_1, s_2, \ldots, s_\alpha) \tag{9.23}$$

and made the additional assumption that

$$f_\alpha(s_1, s_2, \ldots, s_\alpha) = \omega_\alpha[s_1 s_2 \ldots s_\alpha]^{n/\alpha} \tag{9.24}$$

where n, μ_α and ω_α are constants. Equation (9.23) separates the creep function into an elastic and viscoelastic component; the latter is assumed to be a product of functions of its arguments. It is readily seen that (9.23) is not contained in (9.20), and (9.23) provides the elastic component found to be lacking in Stafford's (1969) analysis of Lifshitz and Kolsky's (1967) data.

The particular form of product function (9.24) is motivated by the creep results discussed in Section 1. Each of the terms in (5.1) then produces a contribution

$$\sigma_0^\alpha[\mu_\alpha + \omega_\alpha t^n] \tag{9.25}$$

towards the creep response (9.1) with (9.2). Thus μ_α and ω_α are readily determined by expressing $a(\sigma_0)$ and $b(\sigma_0)$ as polynomials in σ_0. As a test of the theory Findley and Onaran (1968) used the values of μ_α and ω_α determined from tensile single-step creep tests to predict the strain in tensile constant-rate tests. Since the maximum strains attained were only 1%, the distinction between constant rate of loading and constant rate of stressing was ignored, and use of a classical strain measure is justifiable. It was found that predicted and experimental strain values agreed to within approximately 3%.

As noted earlier, Stafford (1969) found that (9.20) gave predictions for two-step creep which differ from experimental data by an apparently elastic contribution. Since (9.23) contains a time-independent component μ_α it might be expected that this assumption should give better predictions of response in two-step creep tests. However, Onaran and Findley (1971) found that this model gave predictions of two-step creep in separate tension and torsion tests with a PVC copolymer which were qualitatively similar to those obtained by Stafford (1969). The product form of constitutive relation gave predictions which were higher than experimental data in step-up tests and lower in step-down tests, but were essentially of the correct time-dependent form. The maximum error within the range of experimental data was approximately 6% of the measured values.

The product form of approximate representation can, of course, also be applied to the stress-relaxation formulation, although available experimental data for stress-relaxation is meagre. Lai and Findley (1968a) discussed a stress-relaxation formulation for uniaxial extension which is completely analogous to (9.23) and (9.24), by considering a relation (5.1) in which Q and P represent σ and ϵ respectively. The material functions Φ_α, which are now stress-relaxation functions, were written in the form (9.23) and (9.24) with the constants μ_α and ω_α determined from single-step relaxation data which could be represented by

$$\sigma(\epsilon_0, t) = \sigma_0(\epsilon_0) - \sigma_1(\epsilon_0)\, t^m \tag{9.26}$$

over the range of time and strain considered. In (9.26) ϵ_0 is the imposed strain level and m is a constant; this expression is the dual of (9.1) and (9.2). It was found that the Lai-Findley theory so obtained was capable of modelling the behaviour of solid polyurethane in various multiple-step uniaxial stress-relaxation tests to within approximately 3%. These tests included strain histories involving increasing and decreasing steps, and appear to indicate that the product form is a satisfactory representation for uniaxial behaviour.

Since it is the stress matrix which appears in the equations of motion, a stress-relaxation formulation may be preferable for the solution of stress-analysis problems. However, as noted previously, creep experiments are simpler to perform than stress-relaxation tests, and it would be a considerable advantage if stress-relaxation functions could be calculated from creep data.

Due to the form of the constitutive relations this is rarely attainable for nonlinear behaviour, but at least one attempt has been made using a number of special assumptions. Lai and Findley (1968b) assumed that their material was weakly nonlinear and that, in a creep test, the time-dependent strain was small compared with the time-independent strain. When these assumptions were coupled with material functions of the form (9.23) with (9.24), they were able to use an iterative procedure to calculate the stress-relaxation functions from creep data. For uniaxial tensile behaviour of polyurethane the functions predicted from creep data agreed reasonably well with data obtained in separate stress-relaxation tests. However, the stress relaxed to only 94% of its instantaneous value during the time scale of the tests and the accuracy of the prediction is only 10% when expressed in terms of the maximum amount of relaxation (that is, 6% of the instantaneous stress). This error was approximately time-independent over the range of data obtained.

In Chapter 5 it was noted that the functions Φ_α can be considered to be symmetric in their arguments *without loss in generality* and, in the subsequent discussion, it has been assumed that this form has been chosen. However, it is not *necessary* that the Φ_α should be taken in a symmetric form, and a different model results from the product assumption if another form is chosen. Without loss in generality, Smart (1972) assumed that $\Phi_n(s_1, s_2, \ldots, s_n)$ is non-zero only when $s_1 \leqslant s_2 \leqslant s_3 \cdots \leqslant s_n$, and the product assumption does not then require that the functional dependences on each s_i should be identical. In terms of symmetric material functions this model is equivalent to expressing the Φ_α as products of arbitrary functions of x_1, x_2, x_3, . . ., where the x_i are equivalent to the s_i arranged in ascending order of magnitude. Smart (1972) considered power-law functions with different indices for each s_i, and compared this and other models with experimental data for polypropylene and PVC, but was unable to detect any general trends in the validity of the models.

B. Three-Dimensional Relations

Product forms for material functions have been used to model data obtained from non-uniaxial tests. Findley and Onaran (1968) considered combined tension and torsion using the creep form of (4.28) relevant to plane stress conditions. By analogy with their one-dimensional assumption (9.24) they assumed that

$$\psi_\alpha(s_1, s_2, \ldots, s_\alpha) = a_\alpha + b_\alpha[s_1 s_2 \ldots s_\alpha]^{n/\alpha} \tag{9.27}$$

where n, a_α and b_α are constants which can be obtained from single-step creep data in tension and torsion by using the method outlined in Chapter 8, Section 2. Expression (9.27) is motivated by its simplicity and by (9.5), but Stafford

(1969) has pointed out that since the ψ_α are not necessarily completely symmetric in their arguments [see (4.30)] this form involves an additional assumption.

As a test of the form (9.27) Findley and Onaran (1968) predicted the behaviour of their PVC in combined tension/torsion experiments involving different time-histories from the single-step creep tests used to evaluate the material data. For constant rate tests their predictions agreed with the experimental data to within approximately 3 %, showing that the model was satisfactory in this situation. A more severe test of the theory was provided by the two simultaneous tension/torsion histories shown in Fig. 1. The Findley-

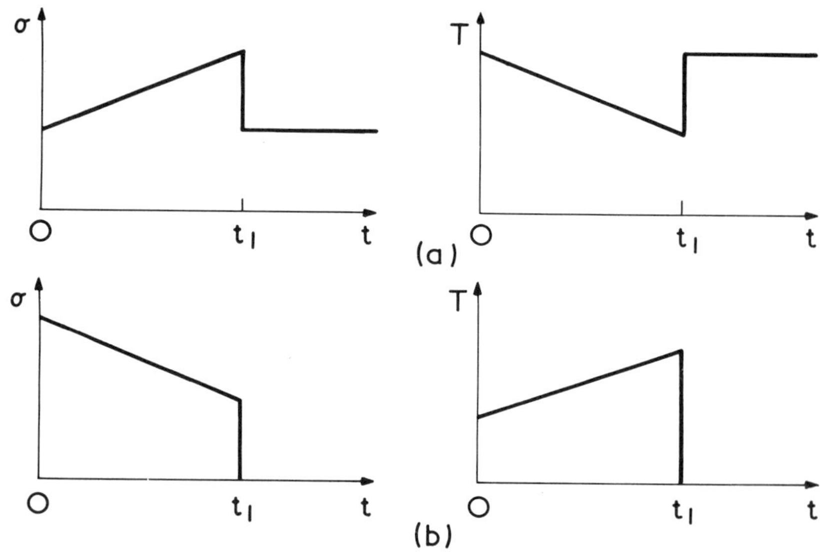

Fig. 1. Loading programmes used by Findley and Onaran (1968) in combined tension (σ)–torsion (T) tests on PVC.

Onaran theory was found to model experimental data for both tensile and shear strains to within approximately 8 % for $t < t_1$. However, for times $t > t_1$ theory and experiment differed by as much as 20 %; in most cases theory tended to overestimate by a significant time-independent amount. Findley and Onaran (1968) claim that this error may be due to ageing of the material or to inadequacy of (9.27). It is possible, however, that this is a manifestation of a more fundamental problem concerning recovery behaviour which is discussed in Section 6.

A further demonstration of the validity of the product form for material functions in single-step situations has been given by Lai and Findley (1969). They considered simultaneous stress-relaxation in tension and creep in torsion for polyurethane, and obtained predictions based on single-step creep data.

The agreement between theory and experiment is good, but the measured tensile stress and creep strain vary by only 10% and 7% respectively during the time-scale of the tests. Thus the material was almost elastic in the experiments which do not, therefore, provide a severe test of (9.27).

Nolte and Findley (1970) considered single-step creep of a material for which volume changes are linearly dependent on the stress history. For this situation a third-order theory requires only seven material functions ψ_α and these are evaluated when each of the arguments is equal to t. These functions can be obtained from two tension, two torsion and one combined tension/torsion creep tests. Nolte and Findley (1970) give a comparison of predicted and meaured creep response for an additional programme of single-step combined tension/torsion tests for polyurethane.

In Section 1C of Chapter 8, it was seen that the viscoelastic response of a number of rubbers can be characterized by a "strain energy" function which is a separable function of time and deformation. In particular, Smith and Frederick (1968) showed that the behaviour of an unfilled styrene–butadiene vulcanizate can be characterized through a single time-dependent function. Valanis and Landel (1967) showed that biaxial and uniaxial relaxation data for a filled dimethyl siloxane rubber could be characterized through a similar relation in which the time-dependent function could be expressed as a simple power. Thus product forms have provided adequate modelling for a number of materials in single-step situations but, in general, the behaviour observed has not been strongly viscoelastic, and conclusions regarding the validity of product forms should be made with caution.

5. OTHER SEMI-EMPIRICAL THEORIES

A. Special Forms of Kernel Function

The product form for the kernel functions ψ_α is just one possible special representation for these multi-argument functions. Other forms involving functions of a single variable, which have been chosen on a purely mathematical basis, have been investigated from the point of view of their physical implications and their ability to model experimental data. Stafford (1969) considered three forms which he named "physical linearity", "superposition" and "product nonlinearity". The model of physical linearity is such that the material functions ψ_i of (4.28) can be expressed in the form

$$\psi_i(s_1, s_2, \ldots, s_n) = \frac{1}{n}\,[f_i^{(1)}(s_1) + f_i^{(2)}(s_2) + \cdots + f_i^{(n)}(s_n)] \tag{9.28}$$

where some of the $f_i^{(\alpha)}$ are related by the symmetry conditions (4.30). This assumption makes the material linearly dependent on the history of the input function, but nonlinearly dependent on the current input. In the stress-relaxation formulation the assumption of physical linearity is directly related to the theory of Coleman and Noll (see Chapter 7, Section 2A) for finite linear viscoelasticity.

In the superposition theory of Stafford (1969) the material functions have the forms

$$\begin{aligned}\psi_i(s_1, s_2) &= f_i(s_1)\, H(s_2 - s_1) + f_i(s_2)\, H(s_1 - s_2)\\ \psi_i(s_1, s_2, s_3) &= g_i(s_1)\, H(s_2 - s_1)\, H(s_3 - s_1) + g_i(s_2)\, H(s_1 - s_2)\, H(s_3 - s_2)\\ &\quad + g_i(s_3)\, H(s_1 - s_3)\, H(s_2 - s_3),\end{aligned} \tag{9.29}$$

where dependence on s_1, s_2, ... has been given the same functional form in order that the ψ_i should be continuous at $s_1 = s_2 = s_3 = \cdots$; this results in completely symmetric expressions.

As noted earlier, Stafford's (1969) product formulation has the material functions

$$\psi_i(s_1, s_2, \ldots, s_n) = f_i^{(1)}(s_1)\, f_i^{(2)}(s_2) \ldots f_i^{(n)}(s_n) \tag{9.30}$$

where certain implications follow from the symmetries (4.30).

The effect of the three types of assumption is most readily demonstrated via a one-dimensional theory. Using superscripts L, S and P to denote the three formulations, equation (5.1) yields

$$Q^L(t) = \int_{-\infty}^{t} \{f_1(t-\tau) + f_2(t-\tau)\, P(t) + f_3(t-\tau)\, P^2(t)\}\, \dot{P}(\tau)\, d\tau \tag{9.31}$$

$$Q^S(t) = \int_{-\infty}^{t} \{f_1(t-\tau) + 2f_2(t-\tau)\, P(\tau) + 3f_3(t-\tau)\, P^2(\tau)\}\, \dot{P}(\tau)\, d\tau \tag{9.32}$$

$$\begin{aligned}Q^P(t) &= \int_{-\infty}^{t} f_1(t-\tau)\, \dot{P}(\tau)\, d\tau + \left[\int_{-\infty}^{t} f_2^{1/2}(t-\tau)\, \dot{P}(\tau)\, d\tau\right]^2\\ &\quad + \left[\int_{-\infty}^{t} f_3^{1/3}(t-\tau)\, \dot{P}(\tau)\, d\tau\right]^3.\end{aligned} \tag{9.33}$$

The forms of the functions chosen in these expressions make the response to a single-step test the same for each theory:

$$Q(t) = P_0 f_1(t) + P_0^2 f_2(t) + P_0^3 f_3(t) \tag{9.34}$$

so that the material functions for all three theories can be determined at one time from single-step data. The relative value of the theories can be investigated by their predictions for multiple-step tests.

Stafford (1969) used the creep and multiple-stage recovery data of Findley and Lai (1966) for unplasticized PVC and of Findley and Stanley (1966) for polyurethane foam to assess his three formulations. For the PVC data it was found that all three theories gave reasonable predictions of the first unloading stage, but the agreement became progressively worse over each subsequent unloading. The predictions of all three theories fell below the measured values, but the superposition form gave the most accurate predictions and the linear form the least accurate. This latter fact implies that the theory of finite linear viscoelasticity (see Chapter 7, Section 2A) may give only a crude representation of material behaviour and, since they are special cases, similar remarks also apply to the theories of Lianis (1963) and McGuirt and Lianis (1967b) [see equations (7.36) and (7.84)]. However, it should be noted that these latter theories are stress-relaxation formulations whereas Stafford's quoted result was for a creep formulation. Partial support for the extension of this result to stress-relaxation formulations is given by Stafford's (1969) analysis of the multiple-step relaxation data of McGuirt and Lianis (1967c) for SBR and of Zapas and Craft (1965) for PIB. Again the superposition formulation gives the most accurate predictions and, in the limited comparisons, is more accurate than the McGuirt-Lianis theory. However, presently available evidence is insufficient to make a definite comparison.

It is noted in passing that the superposition theory is a generalization of Leaderman's equation (9.10). In the special case

$$f_n(s) = \alpha_n f_1(s), \qquad n > 1 \tag{9.35}$$

equation (9.32) can be written

$$Q^S(t) = \int_{-\infty}^{t} f_1(t-\tau)\frac{\partial}{\partial\tau}[P(\tau) + \alpha_2 P^2(\tau) + \alpha_3 P^3(\tau) + \cdots]\,d\tau \tag{9.36}$$

which is equivalent to (9.10) when the stress-relaxation form is chosen.

Gottenberg *et al.* (1969) proposed a different form of special dependence for the material functions in which the ψ_α are equivalent to functions of the sum of their arguments:

$$\psi_\alpha(s_1, s_2, \ldots, s_n) = f_\alpha(s_1 + s_2 + \cdots + s_n). \tag{9.37}$$

When f_α is an exponential function (9.37) is a special case of the product form (9.30) and this form, which was used by Gottenberg *et al.* (1969), leads to (9.33) for uniaxial behaviour. In their experimental programme with a magnetic recording tape material, these authors found that measured and computed loads for single and consecutive constant-strain-rate inputs agreed well initially, but diverged with increasing time.

B. Modified Superposition Theory

In a two-step creep test for which

$$Q(t) = \epsilon(t), \qquad P(\tau) = \sigma(\tau) = \sigma_1 H(\tau) + \sigma_2 H(\tau - k) \tag{9.38}$$

Stafford's superposition theory (9.32) yields

$$\epsilon^s(t) = \epsilon_c(\sigma_1, t) - \epsilon_c(\sigma_1, t-k) + \epsilon_c(\sigma_1 + \sigma_2, t-k) \tag{9.39}$$

where $\epsilon_c(\sigma_1, t)$ is the creep response in a single-step test. Thus the theory requires that each additional load may be treated by superposition as a complete removal of the previous load followed immediately by application of the new total load. This is precisely the motivation used by Findley and Lai (1967) to formulate a modified superposition theory. When ϵ_c is given by (9.1) and (9.2) equation (9.39) gives

$$\epsilon^s(t) = a(\sigma_1 + \sigma_2) + b(\sigma_1)\, t^n + [b(\sigma_1 + \sigma_2) - b(\sigma_1)]\,(t-k)^n \tag{9.40}$$

and, when ϵ^s, σ_1, σ_2, a and b are replaced by three-dimensional matrix forms, this is equivalent to Findley and Lai's (1967) theory. A dual stress-relaxation form is readily derived from (9.32) and has been used by Lai and Findley (1968a) and Stafford (1969). Experimental data of Lai and Findley (1968a) and Onaran and Findley (1971) indicate that the superposition theory gives slightly more accurate results than the product theory. However, Stafford (1969) has shown that this is not true in all cases.

6. RANGE OF VALIDITY OF POLYNOMIAL THEORIES

In Chapter 5 it was shown that the one-dimensional Green-Rivlin constitutive relation (5.1) can be evaluated by performing a set of experiments such as that given in Table 1 of that chapter. The constitutive relation can then be used to predict the response of the material to other loading histories and, as noted for example in Sections 4 and 5, such predictions have not always agreed with experimental data, particularly during recovery. Lockett and Turner (1971) have suggested that this apparent disagreement between theory and experiment is due to misapplication of the theory, and to the fact that a polynomial theory is not the most convenient form. These conclusions arise from the following reasoning.

In a tensile creep test with applied stress σ_0 the creep response $\epsilon_c(\sigma_0, t)$ is given by

$$\epsilon_c(\sigma_0, t) = \sigma_0 \Phi_1(t) + \sigma_0^2 \Phi_2(t, t) + \sigma_0^3 \Phi_3(t, t, t) \tag{9.41}$$

where the series has been terminated after the term of third-order in σ_0. In practice this may be inadequate for some materials and loadings, but the following discussion can be readily extended to any order of integral polynomial. It was noted in Chapter 5 that if three experiments are performed, corresponding to three different values of σ_0 (experiments 1 of Table 1), then the measured creep responses together with (9.41) provide sufficient equations for determining $\Phi_1(t)$, $\Phi_2(t,t)$ and $\Phi_3(t,t,t)$.

In principle (9.41) may now be used to predict the creep response for any other value of σ_0; in particular the creep response in a compressive test is given in the form

$$\epsilon_c(-\sigma_0, t) = -\sigma_0 \Phi_1(t) + \sigma_0^2 \Phi_2(t,t) - \sigma_0^3 \Phi_3(t,t,t). \tag{9.42}$$

Now, the three values of σ_0 used to determine the Φ_α could all have been tensile and it appears, at first sight, that compressive behaviour can be predicted from tensile data. This is unacceptable on physical grounds and, in fact, it leads to a contradiction with experimental results. Using (9.41) and (9.42) it follows that

$$|\epsilon_c(-\sigma_0, t)| = \epsilon_c(\sigma_0, t) - 2\sigma_0^2 \Phi_2(t,t), \qquad (\sigma_0 > 0) \tag{9.43}$$

and since the tensile tests of Lifshitz and Kolsky (1967) (see Chapter 5, Section 3A) and others have shown that $\Phi_2(t,t) < 0$, the theory predicts that

$$|\epsilon_c(-\sigma_0, t)| > \epsilon_c(\sigma_0, t), \qquad (\sigma_0 > 0) \tag{9.44}$$

which is contrary to the experimental fact that materials are never stiffer in tension than in compression.

The fallacy in the above argument is, of course, readily exposed. The Green-Rivlin constitutive relation (5.1) is a polynomial approximation to the general functional relation: similarly (9.41) is a polynomial approximation to the function $\epsilon_c(\sigma_0, t)$. Since there is no reason to suppose that this function is given exactly by (9.41) over the entire range $-\infty \leqslant \sigma_0 \leqslant \infty$, it must be recognized that it is a polynomial approximation and, as such, is necessarily valid only in the range of σ_0 used to evaluate the approximation. Thus it is not permissible to predict compressive behaviour from a model derived from tensile tests. Although this argument is very elementary, it has significant consequences when carried further.

Values of Φ_2 and Φ_3 when their arguments are not all equal can be determined from multiple-step creep tests (see Chapter 5, Section 2). In principle the material functions can be determined from experiments in which each loading step represents an *increase* in the applied stress. By an extension of the argument given above, it does not follow that a model obtained in this manner can predict adequately the response to loading histories involving some unloading steps; this probably accounts for the poor predictions of recovery behaviour obtained from other types of data. When creep functions of several variables are determined through tests involving some positive and some negative steps, the range of applicability of the model becomes very difficult to define.

In order to evaluate the significance of the above remarks it would be advantageous if an experimental investigation were conducted in which monotonic and non-monotonic histories in both tension and compression were considered. This would obviously involve a larger experimental programme than considered previously, but this could help to understand the correlations between load increase, decrease and reversal. Adeyeri *et al.* (1970) have shown that a multiple-integral relation is adequate to model single-step uniaxial creep and stress-relaxation in compression of a remolded clay with a range of water contents. Neis and Sackman (1967) have reported an investigation which goes part way to fulfilling the requirements proposed above. These authors noted that

$$\xi(\sigma, t) \equiv \tfrac{1}{2}[\epsilon(\sigma, t) + \epsilon(-\sigma, t)] = \iint_{-\infty}^{t} \Phi_2(t-\tau_1, t-\tau_2)\, \dot{\sigma}(\tau_1)\, \dot{\sigma}(\tau_2)\, d\tau_{(2)}$$

$$\zeta(\sigma, t) \equiv \tfrac{1}{2}[\epsilon(\sigma, t) - \epsilon(-\sigma, t)] = \int_{-\infty}^{t} \Phi_1(t-\tau_1)\, \dot{\sigma}(\tau_1)\, d\tau_1$$

$$+ \iiint_{-\infty}^{t} \Phi_3(t-\tau_1, t-\tau_2, t-\tau_3)\, \dot{\sigma}(\tau_1)\, \dot{\sigma}(\tau_2)\, \dot{\sigma}(\tau_3)\, d\tau_{(3)} \qquad (9.45)$$

where each of the equations (9.45) can be treated as a special case of (5.1) and the creep functions evaluated accordingly. The response functions ξ and ζ are now simple combinations of the measurements from pairs of experiments which differ only in the sign of the input. This ensures that the model is valid in a region involving both tensile and compressive loads. However, if monotonic loading histories are investigated, it does not necessarily follow that a model based on these data will be adequate to describe non-monotonic loading. For some materials this may be possible, since load reduction can be considered as compression of a deformed material, but investigation of monotonic and non-monotonic histories is required to confirm this. Neis and Sackman (1967) reported experimental data for a low-density polyethylene in various loading histories involving multiple-step loading with load increase, decrease and reversal. Agreement between experimental data and third-order theory is generally good, but the tendency for agreement to be less accurate after load reduction is again evident.

In the above discussion it has been assumed that an integral polynomial approximation is required over a given range of experimental data, and a model is derived which is equally valid over the entire range. In these circumstances the contributions from terms of various orders depend on the range of data and on the order of polynomial chosen (Gradowczyk, 1969). An alternative approach (Drescher and Kwaszczynska, 1970) is to base an approximate description on the concept of regions of linear and nonlinear behaviour. The boundary between the regions depends on the stress state and resembles the boundary between elastic and elastic-plastic regions in plasticity theory. Drescher and Kwaszczynska (1970) show that a smaller number of terms is required in the nonlinear characterization, and give an analysis of their creep data for PVC. Similar separate-regime representations have been considered by Onat (1966) and by Findley and Onaran (1968). These theories are likely to be of greatest value for representing sudden significant nonlinearity for which a single representation would be complex, but application to stress analysis is then more complicated.

In conclusion it is noted that the integral polynomial form of constitutive relation has a limited range of applicability governed by experimental data, and this range may be difficult to define. The existence of a range of applicability is common to all approximate theories but, as will be shown in the next section, formulations can be given for which the definition of this range is simpler than for the polynomial form. In addition these formulations may lead to a reduction in the amount of experimentation required for determination of material functions.

7. OTHER MULTIPLE-INTEGRAL THEORIES

It is readily seen from the discussion of Chapter 5, Section 2, that the factor leading to a heavy experimental requirement is the existence in the constitutive relation of material functions of several variables. If an adequate relation could be determined which combined nonlinearity with material functions of only one, or possibly two, variables, then this would lead to a lighter experimental programme. There are a number of ways in which possible equations of this form may be obtained: examples have been considered in Chapter 7, and also in Sections 4 and 5, where material functions of several time-like variables are expressed in terms of functions of a single variable. Whilst having the desired effect of requiring fewer tests, these theories still suffer from the disadvantages discussed in the previous section relevant to polynomial approximations.

A formulation is now considered in which the dependence on stress is no longer expressed in polynomial form. It is noted that

$$\iiint_{-\infty}^{t} \cdots \int \phi(t-\tau_1, \ldots, t-\tau_n)\, \dot{\sigma}(\tau_1) \ldots \dot{\sigma}(\tau_n)\, d\tau_{(n)} = \int_{-\infty}^{t} \psi(t-\tau)\,[\dot{\sigma}(\tau)]^n\, d\tau \tag{9.46}$$

when ϕ has the form

$$\phi(t-\tau_1,\ldots,t-\tau_n)=\psi(t-\tau_1)\,\delta(\tau_1-\tau_2)\ldots\delta(\tau_1-\tau_n). \tag{9.47}$$

By extracting terms like (9.47) from the higher-order creep functions, and similar contributions with a general dependence on more than one $t-\tau_\alpha$, the creep form of the constitutive relation (5.1) can be rewritten in the form

$$\epsilon(t)=\int_{-\infty}^{t} A[t-\tau,\,\dot\sigma(\tau)]\,d\tau+\iint_{-\infty}^{t} B[t-\tau_1,\,t-\tau_2,\,\dot\sigma(\tau_1)\,\dot\sigma(\tau_2)]\,d\tau_{(2)}+\cdots \tag{9.48}$$

where A and B are new material functions. In their infinite forms (5.1) and (9.48) are equivalent; however, they result in different approximations when they are terminated after a given number of terms. Equation (9.48) has the advantage that it is nonlinear even in its first-term approximation. It is obvious from the derivation that the functions A, B, etc. are not unique in the general infinite expression (9.48); for any truncated series they are to be determined to give the best fit with experimental data. Equation (9.48) has the further advantage that the stress dependence is modelled by functions rather than the simple polynomial terms appearing in (5.1); this simplifies the problem of defining the range of validity of a model.

Alternative forms to (9.48) can be determined in which the stress-rate dependence is replaced by stress dependence or by a mixture of stress and stress-rate dependence [cf. (2.7), (9.31) and (9.32)]; all of these expressions are formally equivalent. One such relation has been suggested by Pipkin and Rogers (1968). Their motivation was that the first integral term should model single-step tests exactly, the second term should extend the exact representation to two-step tests, etc. It should be noted, however, that equations like (9.48) cannot be used as a basis for a *systematic* determination of material properties unless the material happens to be completely definable in one-step tests. This follows since the second term in (9.48) and its alternative forms involves a function of four variables and a systematic experimental programme for its determination would be prohibitive.

The main advantage and purpose of (9.48) is to form the basis for a semi-empirical approach to material representation. As an example, though this is by no means the only possibility, it may be assumed that A, B, etc. are separable in time and stress. Thus constitutive relations are considered of the form

$$\epsilon(t)=\int_{-\infty}^{t}\bar\Phi_1(t-\tau)\,f[\dot\sigma(\tau)]\,d\tau+\iint_{-\infty}^{t}\bar\Phi_2(t-\tau_1,t-\tau_2)\,g[\dot\sigma(\tau_1),\dot\sigma(\tau_2)]\,d\tau_{(2)}+\cdots \tag{9.49}$$

or

$$\epsilon(t) = \int_{-\infty}^{t} \Phi_1^*(t-\tau) \frac{\partial}{\partial \tau}[p\{\sigma(\tau)\}]\, d\tau + \iint_{-\infty}^{t} \Phi_2^*(t-\tau_1, t-\tau_2) \frac{\partial^2}{\partial \tau_1\, \partial \tau_2}[q\{\sigma(\tau_1), \sigma(\tau_2)\}]\, d\tau_{(2)} + \cdots \tag{9.50}$$

In these equations $\bar{\Phi}_\alpha$, Φ_α^*, f, g, p and q are new functions of their indicated arguments. The first-term approximation of (9.50) is similar to Leaderman's equation (9.10) in which the roles of stress and strain have been interchanged.

It is of interest now to consider the predictions of (9.49) and (9.50) for the response to single-step tensile and compressive creep tests. Using a bar and an asterisk to denote the predictions of (9.49) and (9.50) respectively it is found that

$$\left.\begin{aligned} \bar{\epsilon}_c(\sigma_0, t) &= f(\sigma_0)\,\bar{\Phi}_1(t) + g(\sigma_0, \sigma_0)\,\bar{\Phi}_2(t, t) \\ \bar{\epsilon}_c(-\sigma_0, t) &= f(-\sigma_0)\,\bar{\Phi}_1(t) + g(-\sigma_0, -\sigma_0)\,\bar{\Phi}_2(t, t) \end{aligned}\right\} \tag{9.51}$$

$$\left.\begin{aligned} \epsilon_c^*(\sigma_0, t) &= p(\sigma_0)\,\Phi_1^*(t) + q(\sigma_0, \sigma_0)\,\Phi_2^*(t, t) \\ \epsilon_c^*(-\sigma_0, t) &= p(-\sigma_0)\,\Phi_1^*(t) + q(-\sigma_0, -\sigma_0)\,\Phi_2^*(t, t) \end{aligned}\right\} \tag{9.52}$$

where use has been made of the fact that $f(0) = p(0) = g(0,0) = q(0,0) = 0$. It is obvious from these two sets of results that (9.49) and (9.50) have identical forms in this situation.

Since nonlinearity is contained in each of the terms in (9.49), it is reasonable to consider the approximations obtained by considering one or two terms in (9.51) (note that three terms are necessary in the polynomial theory if nonlinear shear behaviour is to be modelled). In the first-term approximation (9.51) shows that compressive behaviour cannot be predicted from tensile data, since they cannot be used to determine $f(-\sigma_0)$. However, to this degree of approximation, the time dependences in tension and compression are identical. In the two-term approximation the time-dependences of the tensile and compressive creep curves are different, and there is again no possibility of misapplying the model to predict the compressive curve entirely from tensile data.

Similar remarks apply to the predictions of (9.49) and (9.50) for two-step tests but, in this case, the two equations do not give identical predictions. In particular, it is readily shown that the responses in a creep-and-recovery test are

$$\bar{\epsilon}_r(\sigma_0, k, t) = \bar{\epsilon}_c(\sigma_0, t) + \bar{\epsilon}_c(-\sigma_0, t-k) + 2g(\sigma_0, -\sigma_0)\,\bar{\Phi}_2(t, t-k) \tag{9.53}$$

$$\epsilon_r^*(\sigma_0, k, t) = \epsilon_c^*(\sigma_0, t) - \epsilon_c^*(\sigma_0, t-k) + 2q(\sigma_0, \sigma_0)\,[\Phi_2^*(t-k, t-k) - \Phi_2^*(t, t-k)] \tag{9.54}$$

and these predictions have interesting differences. In the first-term approximations of (9.49) and (9.50) $\bar{\epsilon}_r$ is given by superposing the tensile creep curve and a suitably translated compressive creep curve; whereas ϵ_r^* is derivable entirely from the tensile creep curve. It is apparent from experimental data that neither of these first-term approximations is adequate in the nonlinear region: in the linear region they are identical and exact. The final terms in (9.53) and (9.54) provide corrections which have been shown by Lockett and Turner (1971) to be adequate to model experimental data of Turner (1966) for a propylene homopolymer. An interesting difference between the two correction terms is that the one for ϵ_r^* requires the function q to be determined only for positive values of its arguments, whereas that for $\bar{\epsilon}_r$ requires evaluation of g for a negative argument. Thus (9.53) treats increasing and decreasing loads as different phenomena, whereas (9.54) implies that the response to one can be obtained from the response to the other. Although the possibility of predicting compressive from tensile behaviour was rejected on the assertion that they are different types of physical behaviour, it is not clear whether unloading and additional loading are basically different. Resolution of this physical question could provide a means of choosing between equations like (9.49) and (9.50).

This section is concluded by noting that (9.49) and (9.50) are merely examples of simple forms of (9.48) which may be postulated for a given material. Other expressions may be more appropriate in practice, and the problem of determining an adequate form is considered in the next section. If no such simple form exists, problems of material characterization and stress analysis for applications requiring high accuracy may be prohibitive.

8. A SEMI-EMPIRICAL APPROACH

In the absence of a rational procedure for simplifying (9.48) by using combinations of functions of one or two variables, one is left with the possibility of using *ad hoc* assumptions similar to those used by Stafford (1969) for the polynomial theory or of using a semi-empirical approach. In the latter the experimental data are modelled by relatively simple mathematical expressions and these are then used to suggest suitable forms for the functions A, B, etc. appearing in (9.48). This is essentially the approach used by Findley and his co-workers, which has been described in Sections 4 and 5, although this work did not relate to (9.48). The first aim of the semi-empirical approach should be to model experimental data for loading histories which are sufficiently complex to introduce a significant degree of nonlinearity and yet are sufficiently simple to allow the implications to be analysed. The results of one- and two-step tests were considered in this context in the previous section; it is necessary now to consider more complex situations.

If triple and higher-order integral terms are necessary in constitutive relations like (9.48) their contributions will be evident in the analysis of multiple-step experiments. A particularly interesting form of this type of experiment, which has been considered by Lockett and Turner (1971), is an intermittent square-wave loading in which a stress σ_0 is alternately applied for a time interval T and removed for an interval S. This form of loading is particularly valuable, since it provides information about the necessity for higher-order terms in (9.48) whilst involving only a single non-zero stress level σ_0. Multiple integral terms characterize nonlinear interaction effects relative to events occurring at different times, whereas simple nonlinearity of response can be included in any of the terms of this model. The particular form of this stress dependence can be considered separately, by performing tests at other stress levels.

As a preliminary investigation Lockett and Turner (1971) considered loading histories for which $S \gg T$; it was presumed that if S/T was sufficiently large the response to each square-wave loading would have become zero before the next load was applied and the responses to each wave would be the same. However, the experimental data for a propylene homopolymer showed that the responses to square-waves were not the same, even though full recovery was attained during the unloaded period. In terms of the Green-Rivlin representation this result implies that material functions $\Phi_\alpha(s_1, s_2, .., s_\alpha)$ have significant values for larger values of s_i when the arguments are not all equal than when they are all equal. In physical terms it shows that interactions between loads may be significant over longer time intervals than direct responses to loads. This fact is relevant to the definition of what constitutes a virgin material: it is not sufficient that residual strains should have vanished.

The phenomenon discussed above is a viscoelastic effect. It does not depend on permanent physical damage to the material since the response to the first cycle was reproduceable if the specimen was left in an unloaded state for a sufficiently long time. It would therefore not be correct to regard the increased response in consecutive cycles as work-softening or some other mechanical conditioning process. This latter interpretation was given to somewhat similar behaviour of the oriented polypropylene specimen used by Ward and Onat (1963). In their tests they used a conditioning process of square-wave cycling before proceeding with the tests described in Chapter 5, Section 3. Since the material possessed a high degree of molecular orientation and was fairly highly crystalline, it is probably fair to consider this as a mechanical conditioning process involving changes to the material structure rather than as a viscoelastic phenomenon. However, the results of Lockett and Turner (1971) for the propylene homopolymer appear to be purely viscoelastic effects.

If $R_n(\sigma_0, t)$ denotes the recovery response of the material in an intermittent loading test following the nth cycle (and $t = 0$ denotes the beginning of the particular recovery phase) it was found that the data of Lockett and Turner

(1971) could be modelled satisfactorily by

$$R_1(\sigma, t) = \eta(\sigma) + \xi(\sigma)\,\theta(t)$$
$$R_n(\sigma, t) = R_1(\sigma, t) + \Sigma(\sigma)\,T(t)\,N(n) \tag{9.55}$$

where, for tests at different stress levels,

$$\xi(\sigma_2)/\xi(\sigma_1) = \Sigma(\sigma_2)/\Sigma(\sigma_1).$$

Similar types of representation could be made of the creep data $C_n(\sigma, t)$ during each cycle. Although these relations are insufficient to determine completely the form of the functions A, B, etc. appearing in (9.48), they appear to suggest that these functions can be expressed as sums of a few separable functions. Evaluation of precise expressions for A, B, etc. would require further experimentation and, in particular, tests for which S is comparable with T.

Benham and Hutchinson (1971) considered square-wave loading in which the material is subjected alternately to tensile and compressive loads. In experiments on PMMA, polyacetal and polypropylene they found that, at intermediate and high cyclic creep stresses, the creep strain increased steadily throughout the test indicating a softening process. For all but the highest stress levels, complete recovery took place within two days after completion of the cyclic programme, which indicated that the behaviour was viscoelastic. Under low cyclic stresses the data were modelled by a superposition theory employing tensile and compressive creep data, but this theory was inadequate to characterize the response at higher stress levels. The problem of modelling these data is similar to that encountered by Lockett and Turner (1971), but a satisfactory model has not been determined. Other data demonstrating the apparent change of stiffness relative to additional loading of a material which has been subjected to different previous loading histories have been given by Turner (1966).

9. CHARACTERIZATION USING REDUCED TIME

A. Constitutive Relations

The linear one-dimensional constitutive relation (2.5) can be rewritten as

$$\sigma(t) = \phi_\infty\,\epsilon(t) + \int_{-\infty}^{t} \Delta\phi(t-\tau)\,\dot{\epsilon}(\tau)\,d\tau \tag{9.56}$$

where ϕ_∞ is the equilibrium, or long time, value of the stress-relaxation function, and

$$\Delta\phi(t) = \phi(t) - \phi_\infty \tag{9.57}$$

denotes the transient component of this function.

Since ϕ_∞ cannot be determined directly from experimental data, the form (9.56) does not hold any practical advantage over (2.5), although, in principle, it expresses stress as the sum of equilibrium and transient components. However, when (9.56) is generalized to nonlinear behaviour, there is no reason to suppose that the nonlinear dependencies of these components is identical, and it may be convenient to consider the equilibrium and transient responses separately.

A formulation which extends (9.56) to nonlinear response has been developed by Schapery (1969a) on the basis of his theory of the thermodynamics of irreversible processes (Schapery, 1965, 1966, 1968). By assuming certain simple forms for free energy and entropy production, he derived the constitutive equation

$$\sigma(t) = h_\infty \phi_\infty \epsilon(t) + h_1 \int_{-\infty}^{t} \Delta\phi(\rho - \rho') \frac{\partial}{\partial \tau}[h_2 \epsilon(\tau)]\, d\tau \tag{9.58}$$

where the reduced time ρ is defined by

$$\rho \equiv \rho(t) = \int_{t_0}^{t} dt'/a_\epsilon[\epsilon(t')], \qquad a_\epsilon > 0 \tag{9.59}$$

where t_0 is an arbitrary time prior to the first non-zero straining of the material and $\rho' = \rho(\tau)$. In (9.58) h_∞ and h_1 are functions of $\epsilon(t)$, where h_2 and a_ϵ are functions of $\epsilon(\tau)$. The strain-dependent properties of h_∞, h_1 and h_2 arise from third and higher order strain effects in the Helmholtz free energy. The strain dependence of a_ϵ arises from similar effects in both entropy production and free energy and may depend also on temperature T, but this will not be considered here. The linear form (9.56) is recovered when $h_\infty = h_1 = h_2 = a_\epsilon = 1$, and these values must be approached as strains become sufficiently small. When

$$h_1 = a_\epsilon = 1, \qquad h_\infty = h_2 = f(\epsilon)/\epsilon \tag{9.60}$$

equation (9.58) reduces to (9.10), and (9.58) can therefore be considered as a generalization of Leaderman's equation. It is also related to other simple equations discussed previously, and these cases are noted later.

The constitutive relation (9.58) has a number of potentially desirable features as judged by the discussion given in Sections 6 and 7. Nonlinearity with respect to strain dependence is contained in the form of functions rather than polynomial terms, so that the relation does not suffer from the interpretational difficulties discussed in Section 6 and it may be adequate to describe strong nonlinearity of some materials. In addition the theory depends on material functions of only one variable, and the required experimental programme is not prohibitive. The only material function depending on a time-like variable is the stress-relaxation function ϕ, and this can be determined in a test at small strain.

Equation (9.58) is a special case of the stress-relaxation form of (9.48) in the sense that it contains only a single integral term and the function A is expressed as a simple sum of products of functions of a single variable. However, (9.58) contains an additional dependence on reduced time which is not included in (9.48). It is not clear that this form of dependence can replace the multiple-integral terms in (9.48), but any justification can come only from the ability of a given equation to model experimental data. Equation (9.58) has some similarity to the equation of Pipkin and Rogers (1968): the former has the extra dependence on reduced time, but a special form for the function A.

Schapery (1969b) has derived a constitutive relation which generalizes the creep formulation of the linear theory:

$$\epsilon(t) = g_0 \psi_0 \sigma(t) + g_1 \int_{-\infty}^{t} \Delta\psi(\gamma - \gamma') \frac{\partial}{\partial \tau}[g_2 \sigma(\tau)]\, d\tau \tag{9.61}$$

where the reduced time γ is defined by

$$\gamma \equiv \gamma(t) = \int_{t_0}^{t} dt'/a_\sigma[\sigma(t')], \qquad a_\sigma > 0 \tag{9.62}$$

where t_0 is an arbitrary time prior to the first non-zero stressing of the material and $\gamma' = \gamma(\tau)$. In (9.61) ψ_0 and $\Delta\psi$ are given by

$$\psi_0 = \psi(0), \qquad \Delta\psi(t) = \psi(t) - \psi_0 \tag{9.63}$$

where $\psi(t)$ is the creep function for small stresses; g_0 and g_1 are functions of $\sigma(t)$, and g_2 and a_σ are functions of $\sigma(\tau)$. The functions g_0, g_1, g_2 and a_σ tend to unity as stresses tend to zero, and (9.61) then reduces to the linear form (2.9). Material behaviour is modelled by five functions of single variables and their experimental determination, which is discussed below, is not a prohibitive task.

B. Single-Step Tests

In a creep test, for which $\sigma(t) = \sigma_0 H(t)$, equation (9.61) gives the response in the form

$$\epsilon_c(\sigma_0, t) = g_0^0 \psi_0 \sigma_0 + g_1^0 g_2^0 \sigma_0 \Delta\psi(t/a_\sigma^0) \tag{9.64}$$

where the superscript denotes that, for example, $g_0^0 = g_0(\sigma_0)$. Equation (9.64) shows that the nonlinear creep compliance ψ_n is given by

$$\psi_n^0 \equiv \epsilon_c(\sigma_0, t)/\sigma_0 = g_0^0 \psi_0 + g_1^0 g_2^0 \Delta\psi(t/a_\sigma^0) \tag{9.65}$$

and this depends on σ_0 in the nonlinear range.

Expression (9.64) is adequate to model the experimental data discussed in Section 1 which was represented by (9.1), (9.2) and (9.3). When time-dependence of the transient term follows a power-law relation, equation (9.65) becomes

$$\psi_n^0 = g_0^0 \psi_0 + \frac{g_1^0 g_2^0}{(a_\sigma^0)^n} \psi_1 t^n \tag{9.66}$$

where ψ_1 is a constant and, when ψ_0, ψ_1 and n are known from tests at small stress, g_0^0 and $g_1^0 g_2^0 (a_\sigma^0)^{-n}$ can be determined, for each σ_0, from straight line plots of ψ_n^0 against t^n. When equation (9.3) applies

$$g_0^0 = \frac{\sinh(\sigma_0/\sigma_a)}{(\sigma_0/\sigma_a)}, \quad \frac{g_1^0 g_2^0}{(a_\sigma^0)^n} = \frac{\sinh(\sigma_0/\sigma_b)}{(\sigma_0/\sigma_b)} \tag{9.67}$$

both of which tend to unity as σ_0 tends to zero. It follows that

$$\psi_0 \sigma_a = \epsilon_a, \qquad \psi_1 \sigma_b = \epsilon_b. \tag{9.68}$$

As noted after equation (9.3) some data can be modelled by a value σ_a such that $\sigma_a \gg \sigma_0$, in which case $g_0^0 \equiv 1$.

When the power-law dependence (9.2) does not apply Schapery (1969a) proposed that material functions should be determined by the following graphical method. From (9.65)

$$\log(\psi_n^0 - g_0^0 \psi_0) = \log g_1^0 g_2^0 + \log \Delta\psi(t/a_\sigma^0) \tag{9.69}$$

and plots of $\log(\psi_n^0 - g_0^0 \psi_0)$ against $\log t$ for different stress levels σ_0 should yield a set of curves which can be superposed by suitable translations in the coordinate directions. If all curves are shifted to the infinitesimal curve, which is strictly valid for $\sigma_0 = 0$, the amounts of shift in the two coordinate directions are $\log g_1^0 g_2^0$ and $\log a_\sigma^0$, and this allows $g_1^0 g_2^0$ and a_σ^0 to be determined.

This method requires prior determination of $g_0^0 \psi_0$ in order to calculate the transient $\psi_n^0 - g_0^0 \psi_0$. Since the instantaneous value $g_0^0 \psi_0$ cannot be measured accurately, Schapery (1969a) proposed that it should be determined by modifying the measured values on a trial-and-error basis so as to obtain curves of $\log(\psi_n^0 - g_0^0 \psi_0)$ against $\log t$ which have the superposition property.

It is evident from the discussion given above that single-step creep tests can be used to determine g_0, ψ, a_σ and $g_1 g_2$ provided that a power-law time dependence does not apply. Separate values for g_1 and g_2 cannot be obtained from single-step data, and must be determined from two-step tests discussed in the next sub-section. When the time-dependence satisfies a power-law relation, that is, when the plots of $\log(\psi_n^0 - g_0^0 \psi_0)$ against $\log t$ are straight lines, g_1^0, g_2^0 and a_σ^0 can be determined only in the combination $g_1^0 g_2^0 (a_\sigma^0)^{-n}$.

For single-step stress-relaxation where $\epsilon(t) = \epsilon_0 H(t)$, equation (9.58) yields

$$\sigma(t) = h_\infty^0 \phi_\infty \epsilon_0 + h_1^0 h_2^0 \epsilon_0 \Delta\phi(t/a_\epsilon^0). \tag{9.70}$$

Schapery (1969a) noted that experimental data of Meredith (1956) and Passaglia and Koppehele (1958) for textile fibres can be modelled by

$$\sigma(t) = \epsilon_0 f(t/a_\epsilon^0) \tag{9.71}$$

and this behaviour is obtained by setting $h_\infty = h_1 h_2 = 1$. In this case the change of strain level merely produces a change in the intrinsic time scale.

When $h_\infty = h_1 h_2$ and $a_\epsilon = 1$ for all ϵ_0 equation (9.70) reduces to

$$\sigma(t) = h_\infty \epsilon_0 \phi(t) \tag{9.72}$$

so that dependence on strain and time is separable. This type of behaviour has been noted in Chapter 8, Section 1C in connection with large-strain stress-relaxation data for rubbers.

C. Two-Step Tests

For a two-step creep test it is convenient here to write the stress history as

$$\sigma(t) = \begin{cases} \sigma_1, & 0 \leqslant t < t_1 \\ \sigma_2, & t_1 \leqslant t \end{cases} \tag{9.73}$$

(Note that σ_2 here denotes the *total* stress for times $t \geqslant t_1$ and not just the increment applied at $t = t_1$.) When (9.73) is substituted into (9.61) the strain for times $t \geqslant t_1$ is given by

$$\epsilon(t) = g_0^2 \psi_0 \sigma_2 + g_1^2 \left\{ g_2^1 \sigma_1 \Delta\psi \left[\frac{t_1}{a_\sigma^1} + \frac{t - t_1}{a_\sigma^2} \right] + (g_2^2 \sigma_2 - g_2^1 \sigma_1) \Delta\psi \left[\frac{t - t_1}{a_\sigma^2} \right] \right\} \tag{9.74}$$

where $g_i^j = g_i(\sigma_j)$ and $a_\sigma^j = a_\sigma(\sigma_j)$.

In a creep-and-recovery test $\sigma_2 \equiv 0$, for which $g_i^2 = a_\sigma^2 = 1$, and (9.74) becomes

$$\epsilon(t) = \sigma_1 g_2^1 \{ \Delta\psi(t_1/a_\sigma^1 + t - t_1) - \Delta\psi(t - t_1) \}. \tag{9.75}$$

When all of the functions g_0, $g_1 g_2$, a_σ and ψ are known from single-step creep tests, recovery data and equation (9.75) allow determination of g_1 and g_2 separately.

By comparison of (9.75) with (9.64), and using the notation of Section 3B, Chapter 5, it can be seen that

$$\epsilon_r(\sigma, t) = \epsilon_c(\sigma, t) - \{\epsilon_c[\sigma, t_1 + (t - t_1) a_\sigma] - \epsilon_c[\sigma, (t - t_1) a_\sigma]\}/g_1 \qquad (9.76)$$

and ϵ_r is given by a suitable translation of ϵ_c only when $g_1 = a_\sigma = 1$. For the arbitrary two-step loading (9.73) it is readily shown that (9.74) yields

$$\epsilon(t) = (g_1^2/g_1^1)\{\epsilon_c[\sigma_1, t_1 + (t - t_1) a_\sigma^1/a_\sigma^2] - \epsilon_c[\sigma_1, (t - t_1) a_\sigma^1/a_\sigma^2]\} + \epsilon_c(\sigma_2, t - t_1). \qquad (9.77)$$

When $g_1^1 = g_1^2$ and $a_\sigma^1 = a_\sigma^2$, equation (9.77) reduces to (9.39) which is the superposition theory employed by Stafford (1969) and Findley and Lai (1967). Equation (9.77) offers a more general superposition law which, by virtue of its extra parameters, may give a better modelling of experimental data. Although no attempt has been made to do so, equation (9.61) may also provide an adequate representation of the intermittent-loading data discussed in Section 7.

D. Experimental Investigations

It has been noted in Sub-section 9B that the constitutive equations (9.58) and (9.61) can be used to model a large amount of existing single-step creep and stress-relaxation data. Schapery (1968) has shown that data for some polymeric materials below their glass-transition temperature can be modelled with the nonlinearity contained entirely in the shift factor a_σ. Setting $g_0 = g_1 = g_2 = 1$ in (9.64) yields an instantaneous response proportional to σ, and (9.67) gives

$$a_\sigma = \left[\frac{\sigma/\sigma_b}{\sinh(\sigma/\sigma_b)}\right]^{1/n} \qquad (9.78)$$

For rubbery polymers above their glass-transition temperature Schapery (1968) noted that the shift factor tends to be insensitive to stress, and reduced time can often be replaced by real time. This model can be used to model much known single-step stress-relaxation data for rubbery materials, and often reduces to the product form (9.72).

Schapery (1969a) has used (9.61) to analyse the tensile creep-and-recovery data of Van Holde (1957) for strips of nitrocellulose film. For short-duration creep ($t_1 = 2.8$ hr) the recovery strain ϵ_r was modelled adequately by a translated creep curve which was seen, following (9.76), to require that $g_1 = a_\sigma = 1$. However, this model was not adequate to represent recovery behaviour in long-duration creep ($t_1 = 12$ days) and it was necessary to introduce a stress-dependent a_σ in that case. This inconsistency is not believed to invalidate (9.61), since the material behaviour was influenced by ageing which is not included in the theoretical model.

Creep-and-recovery data of Martirosyan (1965) for uniaxial tensile behaviour of a phenolic resin with 50% glass-fibre reinforcement was also considered by Schapery (1969a). The reinforcement consisted of two sets of mutually perpendicular fibres and uniaxial loading was considered at angles of 0°, 15°, 30° and 40° to the fibres. In this case it was found that an adequate model was obtained with $g_0 = g_1 = g_2 = 1$, and all of the nonlinearity is due to the function a_σ. For these data the simple linear prediction for ϵ_r was grossly inadequate. Other comparisons between theory and data for multiple-step uniaxial creep of a glass-fibre/epoxy composite is reported by Lou and Schapery (1970, 1971), and the agreement is generally good. These authors also consider multiaxial loading, and some success can be claimed using a theory based on a single mechanical invariant, namely the average octahedral shear stress. This concept was also used by Schapery (1969a) to model the combined tension/torsion single-step data of Onaran and Findley (1965). However, confirmation of the validity of the theory for multiaxial loading is at an early stage, and more information is needed before a definite conclusion can be reached.

It is evident from the discussion given in this section that equations of the form (9.58) and (9.61) hold great promise for the characterization of nonlinear behaviour and, since they contain a number of proposed semi-empirical theories as special cases, they may provide a rational means of choosing between these semi-empirical theories. However, it remains to be demonstrated whether (9.58) and (9.61) have a wide range of validity for different materials and different loading histories. In particular it is not clear whether the existence of a reduced time is always sufficient to permit modelling through a single-integral theory. If this is the case, then a considerable practical simplification results but, otherwise, it may be necessary to consider equations of the form (9.48), possibly combined with the concept of stress or strain-reduced time.

References

Adeyeri, J. B., Krizek, R. J. and Achenbach, J. D. (1970). *Trans. Soc. Rheol.* **14**, 375–392.

Andrade, E. N. de C. (1914). *Proc. Royal Soc., Ser. A* **90**, 329–342.

Andrews, E. H. (1968). "Fracture in Polymers". Oliver and Boyd, London.

Appleby, E. J. and Lee, E. H. (1966). *J. Mech. Phys. Solids* **14**, 375–391.

Arai, T. and Suzuki, I. (1961). *J. Jap. Soc. Testing Materials* **10**, 372–375.

Benbow, J. J. (1956). *Proc. Phys. Soc. London Sect. B* **69**, 885–892.

Benham, P. P. and Hutchinson, S. J. (1971). *Polym. Eng. Sci.* **11**, 335–343.

Bergen, J. T., Messersmith, D. C. and Rivlin R. S. (1960). *J. Appl. Polym. Sci.* **3**, 153–167.

Bernstein, B., Kearsley, E. A. and Zapas, L. J. (1963). *Trans. Soc. Rheol.* **7**, 391–410.

Christensen, R. M. (1971). "Theory of Viscoelasticity". Academic Press, New York and London.

Coleman, B. D., Markovitz, H. and Noll, W. (1966). "Viscometric Flows of Non-Newtonian Fluids". Springer-Verlag, Berlin.

Coleman, B. D. and Noll, W. (1961). *Rev. Mod. Phys.* **33**, 239–249.

DeHoff, P. H. and Lianis, G. (1965). Purdue Univ. Report A and ES 65–6.

DeHoff, P. H., Lianis, G. and Goldberg, W. (1966). *Trans. Soc. Rheol.* **10**, 385–398.

Drescher, A. and Kwaszczynska, K. (1970). *Int. J. Nonlinear Mech.* **5**, 11–22.

Dunn, C. M. R., Mills, W. H. and Turner, S. (1964). *Brit. Plast.* **37**, 386–392.

Eringen, A. C. (1962). "Nonlinear Theory of Continuous Media". McGraw-Hill, New York.

Eringen, A. C. and Suhubi, E. S. (1964). *Int. J. Eng. Sci.* **2**, 189–204; **2**, 389–404.

Ferry, J. D. (1961). "Viscoelastic Properties of Polymers", Wiley, New York.

Findley, W. N. (1960). *SPEJ.* **16**, 192–198.

Findley, W. N. and Khosla, G. (1955). *J. Appl. Phys.* **26**, 821–832.

Findley, W. N. and Lai, J. S. Y. (1966). Brown Univ. Report EMRL-27.

Findley, W. N. and Lai, J. S. Y. (1967). *Trans. Soc. Rheol.* **11**, 361–380.

Findley, W. N. and Onaran, K. (1968). *Trans. Soc. Rheol.* **12**, 217–242.

Findley, W. N. and Stanley, C. A. (1966). Brown Univ. Report EMRL-31.

Fitzgerald, E. R., Grandine, L. D. and Ferry, J. D. (1953). *J. Appl. Phys.* **24**, 650–655.

Flowers, R. J. and Lianis, G. (1969). Purdue Univ. Report A and ES 69–5; also appeared in (1970). *Trans. Soc. Rheol.* **14**, 441–459.

Goldberg, W., Bernstein, B. and Lianis, G. (1966). Purdue Univ. Report A and ES 67–1; also appeared in (1969). *Int. J. Non-Linear Mechanics* **4**, 277–300.

Goldberg, W. and Lianis, G. (1967). Purdue Univ. Report A and ES 67–6; also appeared in (1970). *J. Appl. Mech.*, **37**, 53–60.

Goldberg, W. and Lianis, G. (1968). *J. Appl. Mech.* **35**, 433–440.

Gottenberg, W. G., Bird, J. O. and Agrawal, G. L. (1969). *J. Appl. Mech.* **36**, 558–564.

Gradowczyk, M. H. (1969). *Int. J. Solids Structures* **5**, 873–877.

Green, A. E. and Rivlin, R. S. (1957). *Arch. Rat. Mech. Anal.* **1**, 1–21.

Green, A. E. and Rivlin, R. S. (1960). *Arch. Rat. Mech. Anal.* **4**, 387–404.

Green, A. E. and Rivlin, R. S. (1964). *Arch. Rat. Mech. Anal.* **17**, 113–147.

Green, A. E. and Rivlin, R. S. (1965). *Proc. Royal Soc., Ser. A* **284**, 303–324.
Green, A. E., Rivlin, R. S. and Spencer, A. J. M. (1959). *Arch. Rat. Mech. Anal.* **3**, 82–90.
Green, A. E., Naghdi, P. M. and Rivlin, R. S. (1965). *Int. J. Eng. Sci.* **2**, 611–620.
Gross, B. (1953). "Mathematical Structure of the Theories of Viscoelasticity". Hermann, Paris.
Gurtin, M. E. and Herrera, I. (1965). *Quart. Appl. Math.* **23**, 235–245.
Gurtin, M. E. and Sternberg, E. (1962). *Arch. Ration. Mech. Anal.* **11,** 291–356.
Hadley, D. W. and Ward, I. M. (1965). *J. Mech. Phys. Solids* **13**, 397–411.
Halpin, J. C. (1965). *J. Appl. Phys.* **36**, 2975–2982.
Halpin, J. C. and Pagano, N. J. (1968). *J. Compos. Mater.* **2**, 68–80.
Hayes, M. A. and Morland, L. W. (1969). *Trans. Soc. Rheol.* **13**, 231–240.
Huang, N. C. and Lee, E. H. (1966). *J. Appl. Mech.* **33**, 313–321.
Kolsky, H. (1960). *In* "Proc. Int. Symp. on Stress Wave Propagation in Materials" (N. Davids, ed.), pp. 59–90. Interscience, New York.
Kolsky, H. (1963). "Stress Waves in Solids". Dover, New York.
Kolsky, H. (1966). *In* "Mechanics and Chemistry of Solid Propellants" (A. C. Eringen *et al.*, eds.), pp. 357–379. Pergamon, Oxford.
Lai, J. S. Y. and Findley, W. N. (1968a). *Trans. Soc. Rheol.* **12**, 259–280.
Lai, J. S. Y. and Findley, W. N. (1968b). *Trans. Soc. Rheol.* **12**, 243–257.
Lai, J. S. Y. and Findley, W. N. (1969). *J. Appl. Mech.* **36**, 22–27.
Landau, L. D. and Lifshitz, E. M. (1959). "Fluid Mechanics", Pergamon, Oxford.
Leaderman, H. (1962). *Trans. Soc. Rheol.* **6**, 361–382.
Leaderman, H., McCrackin, F. and Nakada, O. (1963). *Trans. Soc. Rheol.* **7**, 111–123.
Lekhnitskii, S. G. (1963). "Theory of Elasticity of an Anisotropic Elastic Body". Holden-Day, San Francisco.
Lianis, G. (1963). Purdue Univ. Report A and ES 63–11.
Lianis, G. (1967). *Recent Advan. Eng. Sci.*, **3,** 499–512
Lifshitz, J. M. (1964). Brown Univ. Report, Div. Appl. Math., Contract Nonr 562(30)/10.
Lifshitz, J. M. and Kolsky, H. (1967). *Int. J. Solids Structures* **3**, 383–397.
Lockett, F. J. (1965). *Int. J. Eng. Sci.* **3**, 59–75.
Lockett, F. J. (1969). Stanford Univ., Dept. of Appl. Mech. Technical Report No. 195.
Lockett, F. J. and Gurtin, M. E. (1964). Brown Univ., Div. Appl. Math. Report Nonr 562(30)/7.
Lockett, F. J. and Stafford, R. O. (1969). *Int. J. Eng. Sci.* **7**, 917–930.
Lockett, F. J. and Turner, S. (1971). *J. Mech. Phys. Solids* **19**, 201–214.
Lodge, A. S. (1964). "Elastic Liquids". Academic Press, London and New York.
Lou, Y. C. and Schapery, R. A. (1970). Purdue Univ. and Texas A and M Univ. Report AFML-TR-70-113.
Lou, Y. C. and Schapery, R. A. (1971). *J. Compos. Mater.* **5**, 208–234.
Lubliner, J. (1967). *Int. J. Solids Structures* **3**, 513–520.
Martirosyan, M. M. (1965). *Mekh. Polimerov* **1,** 47–54.
McGuirt, C. W. and Lianis, G. (1967a). Purdue Univ. Report A and ES 67–2.
McGuirt, C. W. and Lianis, G. (1967b) Purdue. Univ. Report A and ES 67–8; also appeared in (1970). *Trans. Soc. Rheol.* **14,** 117–134.
McGuirt, C. W. and Lianis, G. (1967c). Purdue Univ. Report A and ES 67–14; also appeared in (1969). "Proc. Fifth. Int. Conf. on Rheology, Kyoto 1968", Univ. Tokyo Press.

McGuirt, C. W. and Lianis, G. (1967d). Paper presented at 38th Annual Meeting of the Society of Rheology.
McGuirt, C. W. and Lianis, G. (1968). Purdue Univ. Report AA and ES 68–2; also appeared in (1969). *Int. J. Engng. Sci.* **7**, 579–599.
Meredith, R. (1956). "The Mechanical Properties of Textile Fibers". Interscience, New York.
Mooney, M. (1940). *J. Appl. Phys.* **11**, 582–592.
Morgan, C. J. and Ward, I. M. (1971). *J. Mech. Phys. Solids* **19**, 165–178.
Morland, L. W. and Lee, E. H. (1960). *Trans. Soc. Rheol.* **4**, 233–263.
Nakada, O. (1960). *J. Phys. Soc. Jap.* **15**, 2280–2288.
Neis, V. V. and Sackman, J. L. (1967). *Trans. Soc. Rheol.* **11**, 307–333.
Nolte, K. G. and Findley, W. N. (1970). *J. Appl. Mech.* **37**, 441–448.
O'Connor, D. G. and Findley, W. N. (1962). (Trans. ASME, Ser. B) *J. Eng. Ind.* **84**, 237–247.
Oldroyd, J. G. (1950). *Proc. Royal Soc., Ser. A* **200**, 523–541.
Onaran, K. and Findley, W. N. (1963). Proc. Joint Int. Conf. on Creep, pp. 2.85–2.97. I. Mech. E., London.
Onaran, K. and Findley, W. N. (1965). *Trans. Soc. Rheol.* **9**, 299–327.
Onaran, K. and Findley, W. N. (1971). *J. Appl. Mech.* **38**, 30–38.
Onat, E. T. (1966). Proc. 5th U.S. Nat. Congr. Appl. Mech. 421–434.
Pearson, J. R. A. (1966). "Mechanical Principles of Polymer Melt Processing". Pergamon, Oxford.
Passaglia, E. and Koppehele, H. P. (1958). *J. Polym. Sci.* **33**, 281–289.
Pipkin, A. C. (1964). *Rev. Mod. Phys.* **36**, 1034–1041.
Pipkin, A. C. and Rivlin, R. S. (1961). *Arch. Rat. Mech. Anal.* **8**, 297–308.
Pipkin, A. C. and Rogers, T. G. (1968). *J. Mech. Phys. Solids* **16**, 59–72.
Riesz, F. and Nagy, B. Sz. (1955). "Functional Analysis". Ungar, New York.
Rivlin, R. S. (1955). *J. Rat. Mech. Anal.* **4**, 681–702.
Rivlin, R. S. (1960). Proc. First Symp. Naval Structural Mechanics, pp. 169–198. Pergamon, Oxford.
Rivlin, R. S. and Ericksen, J. L. (1955). *J. Rat. Mech. Anal.* **4**, 323–425.
Rivlin, R. S. and Saunders, D. W. (1951). *Phil. Trans. Royal Soc., Ser. A* **243**, 251–288.
Schapery, R. A. (1965). Purdue Univ. Report A and ES 65–7.
Schapery, R. A. (1966). *Int. J. Solids Structures* **2**, 407–425.
Schapery, R. A. (1968). Purdue Univ. Report, AA and ES 68–4; also appeared in (1968). Proc. IUTAM Symp., East Kilbride, Springer, New York.
Schapery, R. A. (1969a). *Polym. Eng. Sci.* **9**, 295–310.
Schapery, R. A. (1969b). Purdue Univ Report AA and ES 69–2.
Schwarzl, F. and Staverman, A. J. (1952). *J. Appl. Phys.* **23**, 838–843.
Severs, E. T. (1962). "Rheology of Polymers". Reinhold, New York.
Signiorini, A. (1955). *Ann. Mat. Pura Applicata*, **39**, 147–201.
Smart, J. (1972). Ph.D. Thesis, Imperial College, London.
Smith, T. L. (1958). Jet Propulsion Lab., Memorandum No. 20–176.
Smith, T. L. (1962). *Trans. Soc. Rheol.* **6**, 61–80.
Smith, T. L. and Frederick, J. E. (1968). *Trans. Soc. Rheol.* **12**, 363–396.
Smith, G. F. and Rivlin, R. S. (1957). *Arch. Rat. Mech. Anal.* **1**, 107–112.
Sokolnikoff, I. S. (1956). "Mathematical Theory of Elasticity". McGraw-Hill, New York.
Stafford, R. O. (1969). *J. Mech. Phys. Solids* **17**, 339–358.

Staverman, A. J. and Schwarzl, F. (1956). *In* "Die Physik der Hochpolymeren", Vol. IV (H. A. Stuart. ed.). pp. 126–164. Springer-Verlag, Berlin.
Treloar, L. R. G. (1949). "The Physics of Rubber Elasticity", Oxford.
Truesdell, C. (1965). Proc. 4th Int. Congr. Rheol. 1963. **2**, 3–50.
Truesdell, C. and Noll, W. (1965). "The Non-Linear Field Theories of Mechanics". Handbuch der Physik (S. Flügge, ed.). Springer-Verlag, Berlin.
Turner, S. (1966). *Polym. Eng. Sci.* **6**, 306–316.
Valanis, K. C. and Landel, R. F. (1967). *Trans. Soc. Rheol.* **11**, 243–256.
Van Holde, K. (1957). *J. Polym. Sci.* **24**, 417–427.
Ward, I. M. and Onat, E. T. (1963). *J. Mech. Phys. Solids* **11**, 217–229.
Ward, I. M. and Wolfe, J. M. (1966). *J. Mech. Phys. Solids* **14**, 131–140.
Williams, J. G. and Major, T. J. (1967). *In* Proc. Int. Rubber Conf. pp. 497–509, Maclaren, London.
Wriggles, J. D. and Williams, J. G. (1969). *J. Mech. Eng. Sci.* **11**, 28–39.
Zapas, L. J. and Craft, T. (1965). *J. Res. Nat. Bur. Stand. Sect. A* **69**, 541–546.

Author Index

Numbers in *italics* refer to the page on which the reference is listed.

Achenbach, J. D., 172, *185*
Adeyeri, J. B., 172, *185*
Agrawal, G. L., 169, *185*
Andrade, E. N. de C., 46, *185*
Andrews, E. H., 15, *185*
Appleby, E. J., 154, *185*
Arai, T., 160, *185*

Benbow, J. J., 47, *185*
Benham, P. P., 178, *185*
Bergen, J. T., 144, 145, *185*
Bernstein, B., 113, 114, 116, 119, 120, 125, 133, 136, *185*
Bird, J. O., 169, *185*

Christensen, R. M., 13, 14, 23, 71, *185*
Coleman, B. D., 9, 67, 122, 152, *185*
Craft, T., 120, 121, 169, *188*

DeHoff, P. H., 128, 133, *185*
Drescher, A., 173, *185*
Dunn, C. M. R., 47, *185*

Ericksen, J. L., 68, 69, *187*
Eringen, A. C., 59, 60, 69, *185*

Ferry, J. D., 28, 29, 43, 46, 47, 49, 51, 52, 55, 56, 90, *185*
Findley, W. N., 89, 148, 157, 158, 163, 164, 165, 166, 167, 169, 170, 173, 183, 184, *185*, *186*, *187*
Fitzgerald, E. R., 55, 56, *185*
Flowers, R. J., 136, *185*
Frederick, J. E., 146, 167, *187*

Goldberg, W., 126, 128, 130, 131, 133, 136, *185*
Gottenberg, W. G., 169, *185*
Gradowczyk, M. H., 89, 173, *185*
Grandine, L. D., 55, 56, *185*
Green, A. E., 60, 65, 67, *185*, *186*
Gross, B., 28, *186*
Gurtin, M. E., 25, 33, 72, 91, *186*

Hadley, D. W., 87, 88, 89, *186*
Halpin, J. C., 58, 146, *186*
Hayes, M. A., 58, *186*
Herrera, I., 33, *186*
Huang, N. C., 148, 149, *186*
Hutchinson, S. J., 178, *185*

Kearsley, E. A., 113, 114, 116, 119, 120, 125, *185*
Khosla, G., 157, *185*
Kolsky, H., 29, 45, 49, 50, 51, 78, 83, 84, 85, 91, 98, 163, 171, *186*
Koppehele, H. P., 182, *187*
Krizek, R. J., 172, *185*
Kwaszczynska, K., 173, *185*

Lai, J. S. Y., 89, 164, 165, 166, 169, 170, 183, *185*, *186*
Landau, L. D., 18, *186*
Landel, R. F., 123, 167, *188*
Leaderman, H., 158, 159, 160, 161, 162, *186*
Lee, E. H., 56, 57, 148, 149, 154, *185*, *186*, *187*
Lekhnitskii, S. G., 17, *186*
Lianis, G., 97, 107, 122, 123, 126, 128, 130, 131, 132, 133, 134, 135, 136, 137, 138, 139, 140, 169, *185*, *186*, *187*
Lifshitz, E. M., 18, *186*
Lifshitz, J. M., 78, 83, 84, 85, 91, 98, 163, 171, *186*

Lockett, F. J., 65, 66, 72, 82, 91, 96, 97, 100, 101, 105, 107, 108, 111, 154, 155, 170, 176, 177, 178, *186*
Lodge, A. S., 8, 9, *186*
Lou, Y. C., 184, *186*
Lubliner, J., 150, 151, *186*

McCrackin, F., 161, 162, *186*
McGuirt, C. W., 97, 107, 132, 134, 135, 136, 137, 138, 139, 140, 169, *186*, *187*
Major, T. J., 146, *188*
Markovitz, H., 9, *185*
Martirosyan, M. M., 184, *186*
Meredith, R., 182, *187*
Messersmith, D. C., 144, 145, *185*
Mills, W. H., 47, *185*
Mooney, M., 71, 115, *187*
Morgan, C. J., 90, 91, 137, 139, 140, *187*
Morland, L. W., 56, 57, 58, *186*, *187*

Naghdi, P. M., 60, *186*
Nagy, B, Sz., 21, *187*
Nakada, O., 72, 91, 161, 162, *186*, *187*
Neis, V. V., 172, *187*
Noll, W., 9, 59, 60, 61, 63, 64, 67, 68, 122, 152, *185*, *188*
Nolte, K. G., 167, *187*

O'Connor, D. G., 157, *187*
Oldroyd, J. G., 68, *187*
Onaran, K., 148, 157, 158, 163, 164, 165, 166, 170, 173, 184, *185*, *187*
Onat, E. T., 84, 85, 86, 87, 88, 173, 177, *187*, *188*

Pagano, N. J., 58, *186*
Passaglia, E., 182, *187*
Pearson, J. R. A., 9,*187*
Pipkin, A. C., 98, 99, 152, 174, 180, *187*

Riesz, F., 21, *187*
Rivlin, R. S., 60, 65, 67, 68, 69, 70, 71, 107, 109, 144, 145, 146, 147, 152, 159, *185*, *186*, *187*
Rogers, T. G., 174, 180, *187*

Sackman, J. L., 172, *187*
Saunders, D. W., 107, 146, 159, *187*
Schapery, R. A., 179, 180, 181, 182, 183, 184, *186*, *187*
Schwarzl, F., 55, 161, *187*, *188*
Severs, E. T., 7, *187*
Signiorini, A., 115, *187*
Smart, J., 165, *187*
Smith, G. F., 70, *187*
Smith, T. L., 28, 146, 167, *187*
Sokolnikoff, I. S., 18, *187*
Spencer, A. J. M., 65, *186*
Stafford, R. O., 105, 107, 108, 111, 163, 164, 165, 167, 168, 169, 170, 183, *186*, *187*
Stanley, C. A., 169, *185*
Staverman, A. J., 55, 161, *187*, *188*
Sternberg, E., 25, *186*
Suhubi, E. S., 60, *185*
Suzuki, I., 160, *185*

Treloar, L. R. G., 71, *188*
Truesdell, C., 14, 59, 60, 61 63, 64, 68, *188*
Turner, S., 47, 87, 157, 170, 176, 177, 178, *185*, *186*, *188*

Valanis, K. C., 123, 167, *188*
Van Holde, K., 183, *188*

Ward, I. M., 84, 85, 86, 87, 88, 89, 90, 91, 137, 139, 140, 177, *186*, *187*, *188*
Williams, J. G., 146, 160, *188*
Wolfe, J. M., 87, 88, *188*
Wriggles, J. D., 160, *188*

Zapas, L. J., 113, 114, 116, 119, 120, 121, 125, 169, *185*, *188*

Subject Index

Activation energy, 90
Ageing material, 21, 66, 166, 183
Almost-elastic materials, 154–155
Amorphous structure, 11
Angular momentum, balance of, 16
Anisotropy, 14, 17, 29, 33, 57
Apparent viscosity, 7
Attenuation factor, 45

Balance,
  of angular momentum, 16
  of linear momentum, 16
Beam bending, 37, 47, 49
Biaxial test, 47, 77, 97, 101, 111, 134–137, 140, 145, 167
BKZ Theories, 113–121, 125, 129, 130, 133, 159
  for elastic fluid, 116
  for elastic solid, 114
  modified form, 130, 131
  Signiorini form, 115
Boltzmann superposition principle, 22
Boundary value problems, 13, 33, 57, 154
Brownian motion, 13
Bulk behaviour, 19, 32, 37, 47, 77, 78
  modulus, 18
Butyl rubber, 119, 120, 125

Carbon backbone, 10
Cauchy-Green strain measure,
  left, 64
  right, 61
Chain structure, molecular, 10
Complex compliance, 26, 30, 32, 37, 41, 51, 52, 55, 71, 72, 92
Complex modulus, 25, 30, 32, 37, 41, 42, 48, 49, 52, 55, 71, 72, 91–94, 129
Complex modulus formulation, 25, 71
Composite material, 58, 184
Compressive behaviour, 171, 172, 175, 176, 178
Constant rate test, 40, 121, 132, 136, 137, 139, 164, 166, 169
Constitutive relations, 15, 17
  complex forms, 25–26, 71–73, 129–131
  differential forms, 23, 68
  empirical forms, 14, 156–158
  integral forms, 21, 22, 29, 30, 64–67, 74–140, 148, 150, 153, 158–184
  semi-empirical forms, 158–178
Continuity (mass conservation) relations, 16, 30
Convected stress, 67, 78
Copolymers, 11
Correspondence principle (elastic-viscoelastic analogy), 33, 73
Covalent bond, 10
Creep, 4
  effect of temperature, 55
Creep-and-recovery, 4, 85–88, 90, 121, 160, 162, 175, 182–184
Creep function, 23, 30–32, 37–39, 51, 52
  nonlinear, *see* integral theories
Creep tests, 38, 46, 48
  nonlinear material, *see* integral theories
Crosslinks, 11, 12
Crystalline structure, 11

Damped vibrations, 43, 44
Deformation gradient, 60
  relative, 63
Deformation measure, 60, 61, 116
Deviatoric strain, 31
Deviatoric stress, 31
Differential form of constitutive relation, 23, 68
Dilatant fluid, 6
Dirac delta function, 22
Displacement vector, 15, 62
Dynamic heating, 14, 51
Dynamic tests,
  forced vibration, 41, 49, 50, 92, 130
  free vibration, 43, 44, 49, 50
  pulse propagation, 45, 50
  resonance, 42–44, 49, 50
  wave propagation, 44, 45, 49, 50

Ebonite, 51
Elastic modulus, 17
Elastic-plastic material, 20
Elastic potential, 116
Elastic recoil, 8
Elastic solids, 2, 4, 5, 6, 19, 22, 23, 38–40
almost-, 154–155
classical, 17, 30, 33, 62
finite, 69–71
Mooney, 71, 115, 123, 125
neo-Hookean, 71, 115
nonlinear, 69–71
Elastic-viscoelastic analogy (correspondence principle), 33, 73
Elastomers, 12, 13
Empirical theories, 14, 156–158
semi-, 158–178
Entropy production, 179
Equations of motion, 16
Ethylene propylene rubber, 128
Experimental considerations, 46–51
Experimental programmes and investigations, *see* under individual tests
Exponential extension ratio history, 133, 136
Extension, *see* Tension
Extensional vibrations, 45, 49

Fading memory, 23, 122
Failure, 15
Fibre reinforced material, 58, 184
Finite elasticity, 69–71
Finite linear viscoelasticity, 67, 121–123, 152, 168
Lianis's form of, 122, 133, 138, 159
McGuirt-Lianis form of, 134
Fluids and solids, distinction between 4, 5, 8
Forced vibration, 41, 49, 50, 92, 130
Form factor, 44, 49
Fourier transform, 25
Frame indifference, 61
Frame-indifferent time derivatives, 68
Free energy, 179
Free vibration, 43, 44, 49, 50
Frequency components, 92–94

Glasslike state, 13, 52
Glass transition, 12–14, 51

Green-Rivlin theory, 65–67
accuracy of, 89
linear forms, 66–67
one-dimensional form, 74–94
range of validity, 170–173
relation with Leaderman's theory, 161–162
three-dimensional, 65–67, 95–112

Hamilton-Cayley theorem, 124
Heaviside step function, 38
Helmholtz free energy, 179
Hereditary (non-ageing) materials, 21, 22, 66, 113
Hevea rubber, 51–54
Homopolymers, 11

Incompressibility condition, 16, 98, 99, 100, 114, 115
Incompressible material, 18, 30, 46, 70, 77, 98–112, 114–121, 124–140, 143, 145, 148–151, 153
Integral constitutive relations, 21, 22, 29, 30, 64–67, 74–140, 148, 150, 153, 158–184
Integral polynomial relations,
accuracy, 89
order, 87
range of validity, 170–173
Intermittent square-wave loading, 177–178
Interrelations between material functions, 26–29, 31, 154, 155
approximate, 29
Invariants, 70, 117, 147

Kinematic matrices, 68
Kinetic theory of rubberlike elasticity, 71

Lamé's constants, 18, 30
Leaderman's constitutive relation, 158–161, 169, 175, 179
relation to Green-Rivlin theory, 161–162
Lianis's theory, 122, 159
modified form, 133
non-isothermal form, 138

Linear momentum, balance of, 16
Linear polymer, 11
Logarithmic decrement, 44
Logarithmic stretch-rate history, 137
Long-time behaviour, 4–6, 33, 38–40, 52, 115, 117–119, 123, 134
Loss angle, 25, 41
Loss peak, 53
Loss tangent, 25, 52–54

Mass conservation (continuity) relations, 16, 30
Master curve, 55–56, 139
Material frame indifference, 61
Materials,
  of rate type, 69
  of differential type, 69
Material time derivative, 15
Maxwell model, 34–36, 53–54
McGuirt-Lianis theory, 134
Mechanical conditioning, 177
Memory time, 123, 151
Metals, 2, 3, 9, 12
Molecular chain structure, 10
Molecular orientation, 87, 88, 89, 177
Monomer, 10
Monotonic and non-monotonic histories, 172
Mooney material, 71, 115, 123, 125
Multiple-step tests, 80, 81, 83, 85–88, 90, 96, 97, 102, 103, 112, 121, 132, 137, 160, 162–164, 168–172, 175, 177, 181–184

Natural frequency of vibration, 42–44
Natural reference state, 59
  characterization in absence of, 62
Neo-Hookean material, 71, 115
Newtonian fluids, *see* Viscous fluids
Nitrocellulose, 183
Nominal stress, 46, 48, 159, 160, 162
Nonlinear elasticity, 69–71
Normal stresses, 7, 9

Objective strain measure, 61
Objectivity, 61
Oscillatory tests, *see* Dynamic tests

Phase angle, 41, 131
Plane strain, 108–112
Plane stress, 58, 112, 165
Plastic deformation or flow, 9, 12, 14
Plasticity, 9, 21, 173
Plasticization, 12
Plastics, 1, 2, 9, 10, 12, 13, 14, 51, 141
Poisson's ratio, 18
  complex, 32
Polar decomposition theorem, 60
Polyacetal, 178
Polyethylene, 10, 51, 83, 84, 85, 91, 163, 172
Polyisobutylene (PIB), 55, 56, 119, 120, 121, 125, 133, 169
Polymeric structure, 10
Polymerization, 10
  degree of, 10, 11
Polymethyl methacrylate (PMMA), 51–54, 178
Polypropylene, 84, 85, 87, 90, 91, 139, 165, 176, 177, 178
Polystyrene, 51
Polyurethane, 128, 131, 133, 164, 165, 166, 167, 169
Polyvinyl chloride (PVC), 10, 119, 120, 121, 125, 143, 148, 157, 160, 164, 165, 166, 169, 173
Power law relation, 157, 165, 181
Product form of material functions, 157, 161–167, 174, 183
Pseudoplastic fluid, 6
Pulse propagation, 45, 50

Rate-dependent viscosity, 6, 7, 9
Rate of strain, *see* Strain rate
Recoil, 8
Recovery, 4, 5, 8
  see also, Creep-and-recovery
  and Intermittent square-wave loading
Reduced stress,
  for uniaxial extension, 118
  for biaxial extension, 134
Reduced time, 56–57, 138, 178
Reference state, 15, 59
Relative strain measure, 63
Relaxation function, *see* Stress-relaxation function
Relaxation spectrum, 23, 24

Relaxation time, 23, 24, 35
Repeat unit, 10
Resonance methods, 42, 43, 44, 49, 50
Resonant frequency, 43
Rivlin-Ericksen,
  fluids, 69
  materials, 69
Rotated stress, 67, 78
Rubber, 2, 71, 167, 182, 183
  butyl, 119, 120, 125
  ethylene propylene, 128
  hevea, 51–54
  silicone, 10
  SBR, 126, 132–140, 145
Rubberlike elasticity, 12
Rubberlike state, 13, 52

Shape factor, 44, 49
Shear behaviour, 6, 7, 19, 31, 37, 44, 45, 47, 51, 57, 74, 75, 99–101, 146, 157, 166, 175
Shear-thickening fluid, 6
Shear-thinning fluid, 6
Shift function, 56, 90, 139
Short-time approximation, 148–151
Short-time behaviour, 4–6, 33, 52, 123, 125
Silicone rubbers, 10
Silly putty, 3
Simple fluid, 64
Simple material, 60, 64
Small deformation superposed on large deformation, 127–131, 151–154
Solids and fluids, distinction between, 4, 5, 8
Spring-and-dashpot models, 33–36, 53–54
Standard linear solid, 34–36, 53–54
Storage compliance, 51, 52, 56
Storage modulus, 52
Stored elastic energy, 116
Stored energy functional, 71
Strain energy function, 33, 70, 71, 145, 146, 167
Strain invariants, 70, 117
Strain measure, 59, 76, 158, 160
  objective, 61
  relative, 63
  Cauchy-Green, 61, 64
  classical, 17, 62, 76
Strain rate, 6, 7, 18, 68
  constant, 40, 121, 132, 136, 137, 139, 164, 166, 169
Strain tensor, 17
  deviatoric, 31
Stress,
  convected, 67, 78
  nominal, 46, 48, 159, 160, 162
  nonlinear measure, 159, 160
  normal, 7, 9
  plane, 58, 112, 165
  rotated, 67, 78
Stress pulse, 45, 50
Stress-relaxation, 5
  effect of temperature, 55
Stress-relaxation function, 22–24, 29–32, 37, 39, 40, 52, 56
  nonlinear, *see* integral theories
Stress-relaxation tests, 39, 48
  nonlinear material, *see* integral theories
Stress tensor, 15, 16
  deviatoric, 31
Stretch, pure, 61
Styrene butadiene rubber (SBR), 126, 132–140, 145
Superposition,
  of small deformation on large, 127–131, 151–154
  principle, 22
  theory, 167–169, 178, 183
Symmetry of material functions, 66, 165

Tension/torsion test, 97, 101, 126, 144, 145, 148, 157, 158, 165, 166, 167, 184
Thermal effects, 12–14, 51, 55, 90, 131, 137, 146
Thermodynamic considerations 17, 33, 122, 179
Thermoplastic materials, 11, 12
Thermo-rheologically simple behaviour (Time-temperature equivalence), 14, 55–57, 90, 137, 139, 140, 146
Thermosetting materials, 11, 12
Thermo-viscoelasticity,
  linear theory, 55–57
  nonlinear theory, 137–140
Thixotropy, 7

Time derivatives,
  frame-indifferent, 68
  material, 15
Time-temperature equivalence (Thermo-rheologically simple behaviour), 14, 55–57, 90, 137, 139, 140, 146
Torsional pendulum, 49
Torsional wave, 45
Torsion test, 44, 47–49, 78, 84, 85, 126, 148, 166, 167
Transition region, 13
Triaxial test, 57, 97

Uniaxial tests, 19, 32, 37, 42, 46, 76, 83, 115, 117, 118, 120, 125, 127, 132, 135, 138, 139, 146, 148, 157, 158, 159, 164, 165, 167, 169, 172, 175, 183

Velocity gradients, 68
Velocity vector, 15
Vibrating reed, 49
Viscoelastic fluids, 2, 6–8, 68, 69
Viscoelastic materials, 4, 5, 6, 15, 20, 38, 39, 40
Viscosity, 6, 18
  rate-dependent, 6, 7, 9
  apparent, 7
Viscous fluids, 2, 4, 5, 7, 15, 17, 18, 19, 22, 23, 38, 40, 63
Voigt model, 34–36, 53–54

Wave propagation, 44, 45, 49
Weissenberg effect, 8
Williams-Landel-Ferry (WLF) equation, 55, 90

Yield, 9
Yield stress, 20, 21
Young's modulus, 18, 74
  complex, 32, 129